Diversity in Visualization

Synthesis Lectures on Visualization

Editors
Niklas Elmqvist, *University of Maryland*
David S. Ebert, *Purdue University*

Synthesis Lectures on Visualization publishes 50- to 100-page publications on topics pertaining to scientific visualization, information visualization, and visual analytics. Potential topics include, but are not limited to: scientific, information, and medical visualization; visual analytics, applications of visualization and analysis; mathematical foundations of visualization and analytics; interaction, cognition, and perception related to visualization and analytics; data integration, analysis, and visualization; new applications of visualization and analysis; knowledge discovery management and representation; systems, and evaluation; distributed and collaborative visualization and analysis.

Interaction for Visualization
Christian Tominski
2015

Data Representations, Transformations, and Statistics for Visual Reasoning
Ross Maciejewski
2011

A Guide to Visual Multi-Level Interface Design From Synthesis of Empirical Study
Evidence
Heidi Lam and Tamara Munzner
2010

Diversity in Visualization

Ron Metoyer and Kelly Gaither

ISBN: 978-3-031-01478-9   paperback
ISBN: 978-3-031-02606-5   ebook
ISBN: 978-3-031-00350-9   hardcover

DOI 10.1007/978-3-031-02606-5

A Publication in the Springer series
SYNTHESIS LECTURES ON VISUALIZATION

Lecture #9
Series Editors: Niklas Elmqvist, *University of Maryland*
                        David S. Ebert, *Purdue University*
Series ISSN
Print 2159-516X   Electronic 2159-5178

Interaction for Visualization
Christian Tominski
2015

Data Representations, Transformations, and Statistics for Visual Reasoning
Ross Maciejewski
2011

A Guide to Visual Multi-Level Interface Design From Synthesis of Empirical Study Evidence
Heidi Lam and Tamara Munzner
2010

Diversity in Visualization
Ron Metoyer and Kelly Gaither

ISBN: 978-3-031-01478-9   paperback
ISBN: 978-3-031-02606-5   ebook
ISBN: 978-3-031-00350-9   hardcover

DOI 10.1007/978-3-031-02606-5

A Publication in the Springer series
SYNTHESIS LECTURES ON VISUALIZATION

Lecture #9
Series Editors: Niklas Elmqvist, *University of Maryland*
                David S. Ebert, *Purdue University*
Series ISSN
Print 2159-516X   Electronic 2159-5178

Diversity in Visualization

Ron Metoyer
University of Notre Dame

Kelly Gaither
University of Texas at Austin

SYNTHESIS LECTURES ON VISUALIZATION #9

ABSTRACT

At the 2016 IEEE VIS Conference in Baltimore, Maryland, a panel of experts from the Scientific Visualization (SciVis) community gathered to discuss why the SciVis component of the conference had been shrinking significantly for over a decade. As the panelists concluded and opened the session to questions from the audience, Annie Preston, a Ph.D. student at the University of California, Davis, asked whether the panelists thought diversity or, more specifically, the lack of diversity was a factor.

This comment ignited a lively discussion of diversity: not only its impact on Scientific Visualization, but also its role in the visualization community at large. The goal of this book is to expand and organize the conversation. In particular, this book seeks to frame the diversity and inclusion topic within the Visualization community, illuminate the issues, and serve as a starting point to address how to make this community more diverse and inclusive. This book acknowledges that diversity is a broad topic with many possible meanings. Expanded definitions of diversity that are relevant to the Visualization community and to computing at large are considered. The broader conversation of inclusion and diversity is framed within the broader sociological context in which it must be considered. Solutions to recruit and retain a diverse research community and strategies for supporting inclusion efforts are presented. Additionally, community members present short stories detailing their "non-inclusive" experiences in an effort to facilitate a community-wide conversation surrounding very difficult situations.

It is important to note that this is by no means intended to be a comprehensive, authoritative statement on the topic. Rather, this book is intended to open the conversation and begin to build a framework for diversity and inclusion in this specific research community. While intended for the Visualization community, ideally, this book will provide guidance for any computing community struggling with similar issues and looking for solutions.

KEYWORDS

diversity, inclusion, visualization

Contents

8 Marshalling the Many Facets of Diversity 63

Bernice E. Rogowitz, Alexandra Diehl, Petra Isenberg, Rita Borgo, and Alfie Abdul-Rahman

9 Future of Diversity in Vis 87

Ron Metoyer and Kelly Gaither

Editors' Note

Until this book, all of the entries in Morgan & Claypool's Synthesis Lectures on Visualization series were authored by a single author or a small team of authors. Not so for this book, the first to consist of individual chapters contributed by many authors. When we initially approached Kelly and Ron with the idea, it quickly became apparent that they could not be the sole authors: instead, we decided that as many voices as possible needed to be heard in a book of this nature. The reason is simple: this book is a microcosm of the visualization field as a whole, whether we like it or not, and the onus is on us to take our own advice in order to ensure that the final result is inclusive, diverse, and equitable.

As series editors of these Synthesis Lectures, we are both part of the visualization community, and we were both present in the room and taken aback by the intensity of the "Death of SciVis" panel at IEEE VIS 2016 that Ron and Kelly (and Annie) talk about in the Preface. We attended the diversity panel in Phoenix the following year and were both gripped by the same urge that this conversation must be continued in the community at all costs. We have also heard several personal stories highlighting these issues within computer science and professions and, unfortunately, witnessed and helped resolve some situations ourselves. A book—this book—is just one medium, and it may not even be the ideal one for this topic, but we have been reassured to see several other diversity efforts being launched by like-minded individuals across the visualization field. Let's keep the conversation going!

This is an important book, and we're both immensely proud to see it come to fruition, particularly given that it was finished only a year since the most recent diversity panel. However, please don't confuse pride with ownership: our role as series editors was minimal, and the credit should all go to Kelly, Ron, and the contributing authors. We are all richer for their efforts and hope that *this* book provides an opportunity for more community engagement and action.

Niklas Elmqvist & David S. Ebert
Series Editors, Morgan & Claypool Synthesis Lectures on Visualization

Preface

In the spring of 2016, Bob Laramee and two graduate students were exploring a version of CiteVis [1], a web-based application displaying the number of papers published in each track of the IEEE VIS Conference since its inception. One student observed: "That's interesting. It looks like the SciVis track is getting smaller over time while the other two tracks are growing."

If one charts the number of IEEE VIS publications since 1990, when these data were first recorded, a pattern emerges. Comparing the three VIS tracks—InfoVis, VAST, and SciVis—reveals that SciVis may be in a state of decline. CiteVis shows that, in general, the SciVis track in the conference has grown since its inception in 1990, enjoying a period of expansion for about 12-15 years. The peak of this growth in SciVis, around 2004, coincides with a well-known paper, "On the Death of Visualization" [2]. Since that time, SciVis has been shrinking, while InfoVis and VAST have been expanding.

This observation inspired Bob Laramee to propose a panel for the 2016 IEEE VIS Conference in Baltimore, Maryland, "On the Death of Scientific Visualization." The panel featured several well-known leaders in the field, all of whom had served as SciVis program chairs. A recording of the event (informally called "The Death Panel") can be viewed online.[1] The panelists proposed a range of explanations: the role of cyclical trends; the narrow audience of SciVis, and whether Vis research reaches that audience; a lack of fundamental graphics-related algorithms remaining to be discovered; and problems with the conventions of publishing and conferences in general.

The first question from the audience came from, Annie Preston, a Ph.D. student at the University of California, Davis:

"I didn't hear anyone mention diversity. I think this is a way in which visualization and computer science are not keeping up with other scientific fields. I think in other fields, they are talking more about these issues. And I think if we would like to attract people doing interdisciplinary research that come from a scientific domain, we need to keep up in this respect...what do you think?"

There was a notable silence in the room as the all-male panel deliberated. One panelist noted that he didn't believe there was a diversity problem at all. Other audience members, including Robert Kosara, Erica Yang, Kelly Gaither, and Bernice Rogowitz stepped to the microphone to echo Annie's sentiment and add related thoughts. A lively discussion ensued, most of which was dominated by the question of diversity. At that moment, it became obvious that the visual-

---

[1]Slides: https://youtu.be/ypa1CIrN7eM. Panelists: https://youtu.be/-6TmLPPRjqc.

ization community had reached a critical point and needed to engage in a deeper conversation on this topic.

It was clear, then, that a follow-up panel dedicated solely to the topic of diversity in visualization would be beneficial. In fact, at least two people, Bob Laramee and Kelly Gaither, had this idea. Kelly was so inspired by the discussion that she published a follow-up viewpoints article on this topic [3]. In the spring of 2017, Bob and Kelly independently started to organize follow-up panels dedicated to the topic of diversity in visualization; the IEEE VIS panel chairs informed them of their overlapping ideas, and together they proposed a panel for the 2017 conference.

The Diversity in Visualization panel addressed nuanced and fundamental questions such as: What is diversity? In addition to race, gender, age, and culture, can it refer to subject matter, ideology, and technology use? What benefits might more diversity bring? What factors contribute to the apparent lack of diversity in the Vis community, and what, if any, changes need to be made? What broader lessons can we learn from this conversation? This panel again resulted in lively discussion and engaged even more visualization community members with various points of view. Not only had the dialogue grown, but a critical mass of engaged community members had emerged.

When the visualization community gathered in the "Death Panel" to ask itself why SciVis was waning, its responses mostly reflected existing notions about the field. For decades, "scientific visualization" has meant a narrow expertise, rooted in computer graphics and developed for specific applications. But the answer that received the most excited response was that SciVis needs to engage with more diverse perspectives in order to evolve and grow. Rather than declaring SciVis "dead," the community at large began exploring how to broaden itself. What began was a series of persistent conversations that included people from the community of VIS at large, crossing discipline boundaries and creating meaningful connections across artificial silos of research, background, and perspective.

This book is the continuation of that discussion within the visualization community, bringing together authors and contributors from a diverse portfolio of backgrounds, nationality, race, ethnicity, and gender, all of whom are passionate about increasing diversity and fostering inclusion in the broader field of visualization. Authors were solicited through an open announcement and by choosing thought leaders in diversity and inclusion from within the visualization community. To the best of our knowledge, no book has ever been published dedicated to diversity in visualization. This is a vital and timely theme, touching on the experience and interest of every researcher in visualization. The lessons learned may be applicable not only to visualization, but also to other interdisciplinary, evolving fields, for experienced researchers and newcomers alike. Our goal is to create a foundation for ongoing dialogue, which could help pave the way for accelerated progress in diversity and inclusion.

Annie Preston, Ron Metoyer, and Kelly Gaither
February 2019

Acknowledgments

This book is a collective effort put together by a group passionate about increasing diversity and inclusion in visualization. Without their efforts, we would not have been able to put this book together and we are grateful for their dedication and persistence to put words to the blank page.

We would also like to thank our editors for their out of the box suggestion to embark on this adventure, for their patience while we developed topics and for their time spent reviewing this book. Their support, encouragement, and participation were instrumental in getting to the final product.

Finally, we would like to thank our friends and families for the endless conversations on some tough topics and for their patience while we spent time away from them as we pursued this project.

Ron Metoyer and Kelly Gaither
February 2019

CHAPTER 1

Framing the Conversation

Kelly Gaither and Ron Metoyer

Visualization: The use of computer-supported, interactive, visual representations of data to amplify cognition.

Stuart Card
Jock Mackinlay
Ben Shneiderman

1.1 LOOMING CRISIS

There is a looming global workforce shortage in the computational science and high-tech space, primarily due to a strong disconnect between population demographics and the demographics of those educated to fill these jobs. According to the 2014 United States (U.S.) Census Bureau, there were more than 20 million children under 5 years old living in the U.S., and 50.2% of them were minorities. Current birth/death statistics suggest the U.S. is projected to become a majority minority population by 2040. The U.S. Bureau of Labor Statistics projects 1.4 million computer science-related jobs will be available by 2020, with only 400,000 graduate students qualified to fill them. This leaves a staggering deficit of 1 million unmet high-tech jobs in the U.S. alone. Contributing factors to this shortfall include the following.

- Computer science jobs represent 78% of all science, technology, engineering, and mathematics (STEM) occupations. Only 8% of STEM graduates are in computer science [4].

- Overall, women's representation in computer occupations has declined since the 1990s [5].

- In 2011, 11% of the workforce was Black, while 6% of STEM workers were Black. Although the Hispanic workforce has increased significantly from 3% in 1970 to 15% in 2011, Hispanics were only 7% of the STEM workforce in 2011 [5].

- Native Americans, Pacific Islanders, and Indigenous Peoples have been historically low in STEM employment, registering consistently in the low single-digit percentages [5].

While underrepresented minorities in computer science differ in demographics by country, the McKinsey Global Institute predicts a 2020 global workforce with the requisite college/post-graduate education qualified to fill just 13% of projected labor demands worldwide [6]. India and Brazil are rapidly increasing STEM participation through targeted enrollment programs. Europe, however, is projected to have a similar high-tech personnel shortage as the U.S. [7]. This deficit is much more extreme in emerging countries that are depending on 21st century skill-sets for economic growth.

1.2 MAKING THE CASE: IMPORTANCE OF A DIVERSE AND INCLUSIVE VISUALIZATION COMMUNITY

Visualization taps into the very best capabilities of our brains, transforming data that is fundamentally abstract when presented as numbers into something that communicates and illuminates information ranging from the simple to the complex. Visualization researchers, developers, practitioners, and educators routinely work across traditional discipline boundaries, oftentimes in teams of people that come from a diverse blend of backgrounds, using visualizations as a common language for collaboration. As a community, we are native interdisciplinary thinkers, working at the intersection of science, art, engineering, and technology. By definition, this intersection space is a celebration of diversity, a space in which creativity is allowed to flourish and innovation is key. Are we, however, pushing the diversity and inclusion envelope far enough to begin to address the aforementioned crisis? Is diversity a key to the future of the visualization community, as suggested by Annie Preston?

Innovation Economics is an emerging theory emphasizing entrepreneurship and innovation as key indicators for a thriving economy [8]. Professor Scott Page, in his book, *The Difference*, points out that the ability to see problems differently, not simply being smart, is oftentimes the key to innovative breakthroughs [9]. Thus, the diversity of problem solvers is more important to innovation than any one person's intellectual ability, suggesting that diversity in thought is integral to achieving innovation [10]. Cultivating this diversity of thought necessarily means assembling a multicultural, diverse group of people to work as a team, leveraging what those studying organizational dynamics have known for some time. A team's ability to innovate requires the integration of different perspectives, knowledge, experiences, and backgrounds. This integration or intersection of seemingly unrelated perspectives is crucial to break through creative barriers. In fact, more diverse teams have tangible, measurable benefits to tech innovation. According to a University of Maryland and Columbia Business School joint study, gender diversity at the management level leads to a $42M increase in value of Standard & Poor's (S&P) firms. Additionally, 40% more patents are filed by mixed-sex teams compared to all-male teams [11].

As technology matures, we are seeing an ever closer relationship between society, technology, and science and engineering brought to bear by our increasing need to understand the human condition, prevent human suffering, understand human's impact on our planet, and understand our ever-changing stabilities and instabilities. There is no doubt that the relationship between society, technology, and science is growing more and more dependant on one another. Barnosky et al. defined humanity's grand challenges for science and society as those solving the intertwined problems of human population growth and over-consumption, climate change, pollution, ecosystem destruction, disease spillovers, and extinction [12]. The solution space for these issues necessarily mandates interdisciplinary, collaborative research and development, thus fostering an active exchange of information and ideas.

Dictated by the need to solve larger and more complex problems, there is an increasing need for students and researchers capable of working productively in a multidisciplinary, collaborative environment. It is difficult to imagine working on problems of this magnitude in multidisciplinary teams with multicultural impact without thinking about how to communicate effectively. Visualization allows us to communicate, research, develop, and discover intersections that are key to new insights.

In Richard Gregory's book, *Eye and Brain: The Psychology of Seeing*, he wrote, "We are so familiar with seeing that it takes a leap of imagination to realize that there are problems to be solved [13]." That is, we humans are inherently wired to process and visually assess the world around us. Our ability to see complex problems provides us a unique ability to understand things from a different perspective. It is this inherent ability to think about problem solving visually, particularly in a diverse team, that is the key to breakthrough discovery [14]. Visuals allow us to communicate across cultural boundaries, providing a commonality among all of us—a springboard for communication and collaboration. In *The Shape of Thought*, Clegg and DeVarco state that, "visuals are the language of intuition" [15]. This ability to make sense, to imagine, to communicate across a number of boundaries and barriers lends credence to the prevailing thought that visualization has emerged as our modern-day universal language [16].

As visualization researchers, developers, practitioners, and educators, we sit at the intersection of data, science, engineering, and insight. We are well versed in visual communication, and operate fluidly as bridge builders between disciplines and technology. Our careers have been forged with the knowledge that communication is key, collaboration is vital, interdisciplinary is the future, and problems are getting larger and more complicated. Working in this intersection space provides us with a unique perspective, one that can and should be brought to bear as we broaden and diversify our community moving forward.

1.3   STATE OF THE FIELD

Fundamental to what we do as a community is creativity, without which we would not be successful. It seems hard to imagine that we as visualization practitioners/creatives have not yet realized that our inherent and acquired diversity has placed us in a unique position to address a looming global workforce crisis. Because we work at the intersection, not in spite of, we bring a much-needed perspective to building and maintaining a community that is more in line with our respective national populations. Now is the time to understand what we can and should do, and commit to making diversity and inclusion a priority in our community moving forward.

There are a number of diversity and/or inclusion initiatives implemented to date. The high-tech industry has recognized the need to better understand how to recruit and retain diverse employees. Google, Amazon, Microsoft, among others have diversity programs in place and publish their diversity numbers in an effort to be transparent and pledge their commitment. In 2015, Intel pledged $300M to its "Diversity in Technology" initiative to train and recruit women and underrepresented minorities with the goal of achieving full representation by 2020. Many of the larger academic computing centers count and publish their numbers and have made public commitments to recruiting and retaining a diverse workforce. The visualization community should be no different. We must count and publish our diversity data and pledge our commitment to be more diverse and inclusive. Unfortunately, the visualization community does not officially track gender or ethnic participation in the major conferences. There is no question, however, that women and ethnic minorities (i.e., African American, Hispanic, Native American) are underrepresented in the visualization community. To truly understand and measure the diversity of a community, we must begin to collect this data [20]. It is difficult to understand where to go if we are uncertain about where we are.

1.4 DEFINITIONS AND SCOPE

To this point, we have used the terms "diversity" and "inclusion" somewhat loosely. These terms represent hot button topics in academic and industrial conversations today, especially within the technical fields, yet they are not always clearly defined and or understood. Magurran et al. use a biology-centric definition where diversity is defined as "the variety and abundance of species in a defined unit of study [17]." In this regard, diversity is typically measured in terms of richness and variety where richness is simply the number of species in the unit of study represented out of all possible species, and variability describes evenness in species abundances [17]. Generalizing this definition, a community that is diverse with respect to an attibute exhibits a rich variety of values of that attribute and each of those values is evenly abundant.

Diversity has evolved in organizational literature and generally can be defined as a description of the composition and makeup of a group. While early definitions focused on differences with respect to demographics, perspectives, identity, and culture, more recent definitions include other observable or non-observable characteristics such as functional background, technical differences, or personality [18]. Diversity, therefore, is about differences in a community—with respect to any attribute. It is very much in line with Thomas and Ely's suggestion that diversity describes "the varied perspectives and approaches to work that members of different identity groups bring [19]."

Merging these ideas, we suggest that these different identity groups can cover a broad range of attributes, from ethnicity and gender to research topic and technologies. Therefore, a diverse research community is one in which the many attributes of its members and the research are rich in variety and even in abundance.

Inclusion, on the other hand, has roots in organizational literature but is not determined by the makeup of the community. Rather, it is defined by the culture and values of the community, as it focuses on the degree to which individuals in an organization (or community) feel as if they are a part of important organizational processes [18]. An inclusive community is one that is likely diverse, but more importantly, one that is welcoming to the diverse members of the community and that strives to ensure that all of those members are included in all aspects of the community—especially the organizational processes in which power lies. It is about involvement, opportunity, and empowerment.

With regards to the visualization community, or any academic community, diversity can be defined across a broad spectrum of an individual's attributes including ethnicity, gender, age, and ability as well as a community's attributes such as research content including technologies used, geographical distribution of its members, and research topics. Prompted by the previously mentioned IEEE Visualization panel on Diversity and Inclusion, we have opted in this book to focus on ethnic and gender diversity. We chose this narrow focus because these are the attributes at the center of the diversity and inclusion conversation today. They are the driving and most relevant attributes over which to consider diversity in an effort to directly address the inequities that have not only historically plagued the U.S., but that directly effect populations that are significantly untapped resources in our field, as noted in Section 1.1.

Diversity and inclusion with respect to gender and ethnicity are areas that are ripe for innovation and improvement. Thus, in this chapter and the following three chapters we focus on framing this issue. We discuss the broader sociological context in which the issues exist, the current barriers to achieving diversity and inclusion, and finally, we make the discussion concrete with case studies of real life experiences of members of the visualization community. Chapters 5–7 shift gears to focus on how to achieve a diverse and inclusive community with regards to ethnicity and gender, covering topics of recruitment, retention, and best practices for building an inclusive and diverse community in general.

In Chapter 8, we expand our definition beyond gender and ethnicity to explore several other relevant aspects of diversity that can and will prove important to the visualization community in the years to come. These include diversity of topics, technologies, and geography. Finally, in Chapter 9, we summarize key takeaways from each chapter and discuss the future of the visualization community with regards to diversity and inclusion.

CHAPTER 2

Diversity: A Sociological Perspective

Aviva Frank

To understand the need for diversity and its role in efforts to increase equity, one must first have a sociological framework for understanding how power functions in our society. As sociologists, we push ourselves to make connections between the individual and their society at large. Through sociological inquiry, we are better able to examine the varying levels of meaning embedded within social interaction that are often invisible to the lay-person. By examining ideologies, or systems of belief, we can cut through our surface-level assumptions to explore how they really impact our social reality and everyday lives. We are also more able to see the world from others' perspectives. This helps us understand not only the viewpoints of populations who have been silenced or ignored in our society, but also those of dominant groups.

Western societies tend to be highly individualistically focused, and we view issues like inequality in similarly narrow ways. It is easy for us to think "well, I'm not a bigot, I have never committed a hate crime" and absolve ourselves of responsibility for inequality. "Diversity" as a buzz word continues that individualistic thinking and often lets us believe that the bare minimum is more than enough. Contrary to the conventional wisdom concerning most diversity initiatives, simply including a few people of marginalized identities in a field is not enough. Sociological perspectives resist this kind of individualistic thinking to focus on systems within society and examine how they relate to the individual. Through this lens, diversity is not a goal in and of itself, but a necessary part of our collective efforts to increase social equity.

2.1 STRUCTURAL THINKING

Neither society nor individuals exist independently of one another. Individuals, groups, and institutions all interact with each other to make up our society. Illustrated in Figure 2.1 is a large circle with three concentric circles inside of it: the largest circle represents society, which here means all interactions, relationships, and practices of groups of people. The next circle in represents institutions, the structures that make up society, such as education, the economy, and religion. The circle inside of that represents groups, or collections of individuals held together by shared experiences or identities, such as data visualization scientists, or lesbians, or people with

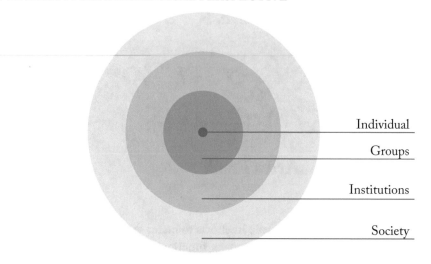

Figure 2.1: The individual exists within a larger social structure.

a college education. Inside of that circle is a tiny dot representing the individual. All of our lives are intimately connected to the groups we belong to, the institutions that we interact with, and the society we live in. While we are all individuals and possess autonomy, none of our actions and opinions are free of social influences. A sociological approach to society and its institutions allows us to see beyond how institutions are purported to work to examine how they actually function in our society at large and in our individual lives. For example, education is not only used to transmit knowledge to younger generations, it is also used to teach children the rules (both explicit and implicit) of society, including a respect for authority and the behaviors and qualities that will make them obedient and productive workers later in life.

Sociologists view norms, values, beliefs, and language as part of a social system that is learned and perpetuated through repetition. Critical engagement with such norms helps us de-naturalize social behaviors that often go unquestioned in society. Some of our most interesting inquiries stem from asking why these social systems exist. For example, if we argue that the norm of paying women less than men, especially women of color in comparison to white men, is not due to an innate biological inferiority, but due to social norms, we are better able to explore questions concerning the history of these systems and how they work. Who benefits from women being less financially independent? Why would women's labor be less valued in society? What other instances do we see in society that have similarly oppressive impacts on women? How do institutions like capitalism, education, and family and social institutions like race, class, and ability interrelate with this phenomenon? With the knowledge we glean from these questions, we can then begin to disrupt and dismantle these oppressive systems in the hopes of pushing society toward a more equitable paradigm.

2.2   SOCIALIZATION

Through the process of socialization we internalize the social norms, attitudes, and values of our society. Put simply, through socialization, we learn the "rules" of how to exist in our society. Those who transgress these rules, or appear to have transgressed them, are negatively sanctioned or punished while those who adhere to, or appear to adhere to, the rules are positively sanctioned or rewarded. These norms are a mechanism of social control. By learning what is acceptable, and realizing that there are punishments for violations and rewards for adherence, we are trained to police ourselves, and others, into conformity.[1]

Nearly from the moment that we are born, we internalize a myriad of cultural mores and ideas about what is acceptable, what is normal, and what is expected of us. This process is important as it teaches us how to be members of a society, however, no society is free of inequality. Many of our social rules and norms function in ways that are oppressive, and as socialized members of society, we perpetuate them, often unconsciously. Just like society shapes us, we shape society, and our repetition of the norms we have been inculcated into throughout our lives further reinforces them as normal and right. When we are socialized in a society in which cultural norms reinforce and tacitly (and sometimes explicitly) approve of systemic bias in favor of certain groups at the expense of others, we perpetuate those biases and reproduce inequality. Furthermore, because institutions are made up of people and designed by people, all of whom have been socialized within these oppressive paradigms, inequality is built into the structures of our society. This process is recursive, or self-reproducing, but also constantly shifting as norms change and people work to change them.

2.3   OPPRESSION

Oppression, too, is a structural issue and works on multiple levels to systematically disadvantage members of certain groups. Social stratification divides society into hierarchized groups, leading to the unequal distribution of wealth, power, access, respect, and rights. The basic formula for oppression is prejudice plus power. For example, as illustrated in Figure 2.2, prejudice around race plus power creates racist oppression.

To be considered oppression, actions, beliefs, and norms need to not only have the prejudices that we have internalized throughout our lives, those prejudices need to be backed up with institutional or systemic power. These could be racist, sexist, classist, or ableist oppressions. For instance, a woman telling a friend that she hates men is not oppressive because our social hierarchies value men and masculinity over women and femininity, whereas a man telling a friend that he hates women is a form of oppression because his opinion mirrors the larger social system that

---

[1]French philosopher and critic Michel Foucault argued that social control, or power, is omnipresent and not only acts on all individuals but is enacted by all individuals. In this way, all individuals are made both the object and the instrument of this "domination-observation." We are sensitized to deviance from the norm and self-surveil and surveil others to limit and punish deviance in order to maintain norms. This process naturalizes norms and constitutes norm-breakers as deviant, and therefore in need of correction or punishment [22].

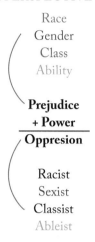

Race
Gender
Class
Ability

Prejudice
+ Power
Oppresion

Racist
Sexist
Classist
Ableist

Figure 2.2: The anatomy of oppression.

enforces gendered inequality. No opinion exists in a vacuum, and in this example, the woman's animosity toward men stems from systematic oppression of women, whereas the man's animosity stems from an internalization of sexist norms and a belief in his superiority. Oppression isn't a "you" thing, it's an "all of us" thing. We all participate in structural oppression, because we all participate in society.

At the end of the day, all of this boils down to power.[2] Dominant groups, for example white, able-bodied, or heterosexual people, get tacit benefits that are imbued from society. Members of these groups get opportunities that are just not accessible to others, and greater access to rights and resources while those who experience the brunt of intersecting oppressions have less access and political power. Members of a dominant group benefit from the oppression of other groups through heightened privileges relative to others—greater access to rights and resources, and better quality of life. This is not to say that white or heterosexual people do not experience suffering, but they don't experience it because of their race or sexuality. Those who are oppressed have less access to rights and resources than those in the dominant group(s), experiencing less political power, lower economic potential, poorer health, and higher mortality rates.[3] This institutional oppression is built into our society, and is the bedrock of our economy,

[2]In his book *The History of Sexuality*, Foucault argues that power is "...produced from one moment to the next... Power is everywhere; not because it embraces everything, but because it comes from everywhere. And 'Power,' insofar as it is permanent, repetitious, inert, and self-reproducing, is simply the over-all effect that emerges from all these mobilities" [23]. Power is not a singularly top-down phenomenon, it is produced everywhere and is present in all interactions.
[3]Iris Marion Young's seminal essay "Five Faces of Oppression" [25] divided oppression into five types, or faces: exploitation, marginalization, powerlessness, cultural domination, and violence. Young's schema enables us to conceptualize the varied ways oppressive power systems structure our daily lives, institutions, and emotional landscapes. Not all oppressed groups experience all five faces of oppression, and certainly not in equal amounts. This flexibility allows us to place different forms of oppression that are often do not easily fit into the oppressed privilege dichotomy into a larger framework of power without losing nuance.

our media, our education system, our medical system, our government, our religions, and our laws. But even more so, these oppressions are built into our social consciousness. The norms and values we are socialized into and the practices we see as normal also perpetuate, and support, institutionalized oppression. This too is recursive. All systems are made up of individuals and all individuals have been socialized in a society that is inequitable. Through socialization we internalize these inequitable ideas and then recreate them in individual and institutional ways. The recreation of inequality normalizes and further codifies inequality in society, leading to it being socialized into others.

One of the insidious characteristics of institutional oppression is the way that systems of inequality interact together, reinforce each other, and render each other invisible. Theorist and philosopher Marilyn Frye described this phenomenon through the metaphor of a birdcage.

> Consider a birdcage. If you look very closely at just one wire in the cage, you cannot see the other wires... There is no physical property of any one wire, nothing that the closest scrutiny could discover, that will reveal how a bird could be inhibited or harmed by it except in the most accidental way. It is only when you step back, stop looking at the wires one by one, microscopically, and take a macroscopic view of the whole cage, that you can see why the bird does not go anywhere... the bird is surrounded by a network of systematically related barriers, no one of which would be the least hindrance to its flight, but which, by their relations to each other, are as confining as the solid walls of a dungeon [24].

The more we study these social systems, the more we see how interconnected and interdependent they are emphasizing how deeply embedded they are in the bedrock of our society. Much of these oppressions happen quietly within institutions that are seen as neutral or natural. For example, the rule of law is purported to carry out justice that is blind to factors like race or class, but in actuality, a belief in the purity of rule of law masks the deep disparities that are knit into the fabric of our laws.[4] Or, consider data—embedded in the concept is the assumption of truthfulness and impartiality that hides the implicit biases of those who created the experiment, collected the data, interpreted it, and represented it [26]. If you take anything from this chapter, perhaps the most important lesson is that nothing is outside of systems of inequality, no matter how harmless or neutral these systems seem.

Generally, it is easier to identify individual discrimination than institutional discrimination. We know that treating someone worse because of their race or sexuality or ability is wrong, and we can often recognize it when we see it. What we often have trouble seeing is the institutional discrimination, in which institutions systematically discriminate against members of a marginalized group. For example, the disproportionately high rates of mass incarceration of

---

[4]Critical Race Theory is a field of legal studies that explores the embedded nature of white supremacy in the U.S. legal system through the use of a critical understanding of history, a disruption of mythologies that obscure the racism built into our institutions, and the elevation of counter-hegemonic voices. Critical Race theorists assert that the legal system is far from neutral and in fact perpetuates racial inequality.

Black Americans, or the disproportionate rates of deportation of non-white undocumented immigrants in comparison to that of white undocumented immigrants. The reproduction of social inequality is the process through which institutional norms of discrimination, which are based on codified prejudicial norms, lead to social disadvantage which then reinforces the prejudice once again. For example, racist ideology that paints Black men as violent and uneducated is prominent in our cultural environment, these beliefs lead to institutional discrimination against Black men such as disproportionately high rates of murder by police officers, incarceration, and job insecurity, which then reinforces the prejudices that underlie that discrimination [27]. As members of an unjust society, our social institutions and our behaviors and norms all perpetuate inequality, whether or not the individuals participating intend to.

2.4   PRIVILEGE/DISPRIVILEGE

In Peggy McIntosh's seminal essay "White Privilege: Unpacking the Invisible Knapsack," she describes white privilege as "...an invisible weightless backpack of special provisions, maps, passports, codebooks, visas, clothes, tools and blank checks [28]." This is true for all institutional privilege, just by existing, as a person with privilege, we are equipped with more access and power than those who have less privilege. Systems of inequality are so neatly woven into the fabric of our society that their influence is often concealed, leading to an unconscious perpetuation of oppression simply through participation in these institutions. Because dominant ideologies are created by dominant groups, we often don't see these privileges and benefits that we are carrying around. We all have differing privileges and disprivileges that shape the way that we interact in everyday life. However, it is not nearly as simple as "privileged" vs. "disprivileged" because none of us are any one thing, and no social category is monolithic.

Multiple axes of identity interact on multiple and often simultaneous levels to create an individual's lived experience. Axes of power, like gender, class, ability, and race, all intersect to create one's own specific life.[5] Not all members of a group are homogenous, and each person's experience of that identity is informed and influenced by their other identities. Some of this is due to differences in individuals' experiences and environments, but a lot of this difference is a result of the individuals' axes of identity and oppression. This also complicates the privilege vs. disprivilege conversation because oppression is experienced differently depending on multitudes of factors. For example, a Black American woman has very different experiences of racism, sexism, and racialized sexism than a white woman, or a non-American Black woman, or an American Black man. Our identities don't simply add or subtract privilege, they interact and inform one another to create complex social positionalities that extend into all aspects of our lives. Imagine a paint palette where each color of paint represents an axis of identity, red

---

[5]The term "intersectionality" was first coined in 1989 by Kimberlé Crenshaw [29] in her paper "Demarginalizing the Intersection of Race and Sex: A Black Feminist Critique of Antidiscrimination Doctrine, Feminist Theory and Antiracist Politics." Crenshaw, a founder and leading scholar of Critical Race Theory, used the term to describe the unique oppression of Black women in the U.S. and U.S. legal precedent. Since then, the term has into a broader social theory that explores the interrelatedness and compounding effect of axes of oppression.

for class, blue for race, yellow for gender, etc. An additive model of oppression would imagine a person's experience of social hierarchization represented as distinct dollops of paint, the person experiences class, race, and gender-based privilege or disprivilege individually. An intersectional model would mix the paints together to create a unique shade that incorporates all axes of identity, arguing that a person's experiences of class, race, and gender cannot be separated, and that their identities actually change how they experience their other identities.

2.5   VISUALIZATION: A CASE STUDY

Looking at the makeup of the data visualization community can help elucidate how complex and insidious institutional oppression can be. Numerous studies have shown that in Western countries, girls in elementary and middle school have similar levels of interest and confidence in their abilities in STEM fields as boys their age, but in high school, these numbers drop precipitously. This is credited to the social messages that equate science and math with masculinity that make the fields inhospitable to girls and women [30]. This curtails the number of women who enter collegiate and graduate programs in science and math, and sexism and harassment within institutions drives even more women away [31]. This leads to a pool of candidates for jobs in the field of data visualization that skews toward men. This gap is even wider for women of color in STEM fields [32]. Further, people of color, due to institutional racism and historical economic oppression, tend to be less wealthy and subsequently, live in areas with poorer school systems. This leads to Latinx[6] and Black children, particularly Latinx and Black girls, being undereducated in math and science. Furthermore, due to soaring higher education costs and the negative impacts of poor school systems, low numbers of Latinx and Black students are achieving higher education and careers in math and science fields [33]. An onlooker may see the lack of women, people from poorer backgrounds, and people of color trying to enter data visualization jobs as indicative of a lack of interest or ability in science and math, but a sociologist sees it as a direct result of historical and institutional oppressions.

Furthermore, our internalized biases have a large effect on how we treat others, the language we use, and the decisions we make [34]. Have you ever seen a woman colleague's ideas be ignored or devalued in favor of a man's similar or worse idea? Have you ever seen a person of color in your lab or office and briefly had the thought "what are they doing here?" or "they're probably only here because of affirmative action." Have you ever wondered about the capability of colleagues with disabilities? Have you ever seen someone be given a raise or promotion over a candidate who was more qualified or deserving, but who was a woman, or a person of color, or had a disability? You probably have. These instances are not anomalies, nor are they isolated events of sexism or racism or ableism—they are a product of institutional social hierarchies. It would be a mistake to assume any field or social arena is free of these biases, regardless of how diverse it seems, or educated its population, or liberal its politics.

---

[6]Latinx is a gender-neutral term for people of Latin American culture and descent.

2.6   NEXT STEPS

In the face of the bleak reality of the omnipresence of oppression I remain hopeful. Perhaps it is naïve, but I don't think you can be a good sociologist without a healthy dose of hope. It's our job to study the many and varied ways people are cruel to one another, and without the hope that the knowledge we glean through our studies could push back against this tide of suffering, we would surely break under its weight. We know from study after study that the more people come into contact with others unlike them, the less prejudiced they tend to be [35]. That, in and of itself, is justification for increased efforts toward diversity. Diversity makes harboring prejudice harder.

Diversity is also a tool for harm reduction. If those in positions of privilege have trouble identifying when oppression happens, then a population made up primarily of those with privilege is more likely to perpetuate inequality than one with people of varying privileges and disprivileges. By making environments more welcoming and accessible to more people—the essence of diversity—we bring more voices into the conversation. Furthermore, simply talking about these issues, bringing them into the light and examining our roles in them, can be productive in and of itself. By acknowledging and understanding these power structures, we can work to do better. Who better to speak on the experiences of the oppressed than those who experience oppression? And who better to direct us toward a more equitable paradigm than those who have fought under the current inequality of this one? By listening to their voices, amplifying their ideas, and learning from their experiences, we can learn how to be less complicit in systems of oppression. Diversity is an essential stepping stone toward a more equitable society for us all.

CHAPTER 3

Factors Hindering Diversity

Kelly Gaither

Over the last several years, we have seen tremendous efforts to try and bring diversity and inclusion to the forefront of everyone's minds. Corporations such as Google, Amazon, Intel, and many other Fortune 500 companies have started initiatives to ensure that they integrate inclusion and diversity into every aspect of their business. Creating a more diverse organization is relatively straightforward. That isn't to suggest that building a truly diverse team of people is easy. Rather, building a diverse team is quantitative and tangible in nature. It is about numbers. Inclusion, on the other hand, is not. It is qualitative in nature and comprises a more complex set of issues that are at times tough to pinpoint, and in many cases even harder to get right. Those working in this area know that it can, at times, feel as if we are climbing over an insurmountable wall, with every one step forward, we take two steps back. Perhaps this is indicative of our efforts in this space being in its infancy. Perhaps it is because, while we all, in theory, agree that these efforts are good and should be pursued, most of us aren't familiar with the vast body of research that informs motivation, best practices, and pitfalls. In this chapter, we provide an overview of the motivation and broad strokes at issues hindering inclusion and diversity.

3.1   THE STRAIGHT LINE FROM DIVERSITY AND INCLUSION TO INNOVATION

We mentioned previously that more diverse teams are more creative and that this creativity and innovation very often lead to positive economic outcomes. As demonstrated by examining the financial returns of Standard & Poor's Top companies for diversity, effectively managing diversity in the workforce is critical to competitiveness in a global market. Scott Page, a Professor of Complex Systems, Political Science, and Economics at the University of Michigan and the author of *The Difference: How the Power of Diversity Creates Better Groups, Firms, Schools, and Societies*, found that progress and innovation depend less on the perceived intelligence or IQ of individual thinkers and more on diversity of perspectives and thought from a group of individuals able to work together and share ideas [9]. He showed that groups with a range of perspectives consistently outperform groups of like-minded experts. Kerby and Burns examined the macro level benefits and delineated the top economic facts of diversity in the workplace, including the following [37].

- A diverse workforce drives economic growth by bringing together varied perspectives that often result in greater innovation.

- A diverse workforce is more able to capture a greater share of the consumer market by appealing to a larger fraction of the population.

- A diverse and inclusive workforce helps businesses avoid employee turnover costs by fostering an environment that welcomes and values employees from all backgrounds.

- Diversity fosters a more creative and innovative workforce by encouraging all perspectives.

- Diversity in the boardroom is needed to leverage a company's full potential and becomes a self-fulfilling prophecy as younger generations see people in leadership positions that look like them.

As noted in Chapter 1, we can consider diversity from three separate but interrelated perspectives, such as demographics, and organizational and socio-cognitive factors [38]. As a reminder, demographic diversity is most easily characterized by characteristics including, but not limited to, age, gender, ethnicity, and nationality. Organizational diversity adds perspectives, group dynamics, education, occupation, seniority, and hierarchy within teams and other organizational hierarchies. Socio-cognitive diversity includes factors relating to cultural and religious values, beliefs, expertise, and personality. Choy illustrates the interconnected nature of all these, what he calls the construct domain of diversity, as shown in Figure 3.1.

It is worth noting that while we are currently talking about diversity with respect to business and economic competitiveness, the concepts we discuss are relevant to all organizations, including higher education and by extension academic competitiveness. These issues are relevant to all types of organizations and teams. Ensuring organizational and team diversity is a constant challenge, a continuous process that evolves and adapts to changing demographics, priorities, and responsibilities. For all organizations, examining diversity at the macro level includes a careful look at policies and formal education programs that create awareness and social consciousness and stresses the importance of culture, thereby ensuring responsibility at all levels of accountability. Diversity at this level includes buy in from members of the organization that comes with empowerment, thus giving members the ability to actively participate in initiatives and ensuring that diversity is institutionalized. This necessarily requires a careful eye toward encouraging thought leaders from a diverse set of sociocultural backgrounds. Examining diversity at the micro-level ensures an eye on group dynamics to maintain diverse thought and approaches in teams, ensuring that all voices are heard and respected, thereby fostering a sense of importance and inclusions in the teams themselves. While much of this may seem like common sense, in practice, ensuring that your organization and teams are diverse and inclusive requires attention and careful consideration and is significantly more difficult to implement in practice than it is to write about in theory. However, we can provide best practices backed by sound research to distill those issues that are most insidious in hindering what we all seek—diverse and inclusive

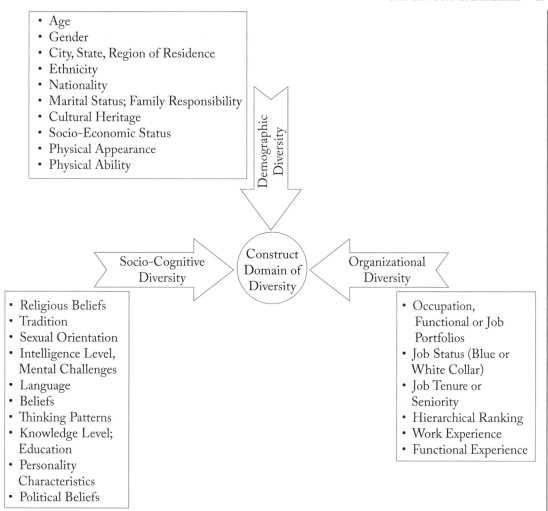

Figure 3.1: Choy's construct domain of diversity.

teams and organizations. What follows is a brief, but useful, overview of the high-level issues that hinder all that wish to create and foster diverse and inclusive teams and organizations.

3.2   ENTRY BARRIERS

In their paper "Perceived Barriers to Higher Education in STEM among High-Achieving Underrepresented High School Students of Color," Scott and Martin identify four types of barriers that affect entry to STEM and higher education from underrepresented groups [41]. These four types of barriers include the following.

- **Structural**—Defined as barriers beyond a person's individual control that exist materially part of the institution or environment. Critical Race Theory suggests that structural barriers to entry in STEM are a consequence of long-term institutional and structural racism in schools and society. These barriers most often impact low-income students and students of color resulting from a lack of access to school funding, science resources, high-quality teachers, technology, and advanced coursework [42–48]. At the undergraduate level, these structural barriers often include lack of sufficient high-school preparation [49], lack of mentorship from role models that "look" like them [50], and perception of an unwelcome and in some cases hostile environment [51–53].

- **Social/Psychological**—Underrepresented students also face social/psychological barriers, as a result of coping with the cumulative effects of stigma and marginalization. These social/psychological barriers can be summarized into four categories [54]: (1) negative treatment and effects of direct discrimination; (2) self-fulfilling prophecies, stereotype threat, under-performance, and disengagement; (3) automatic stereotype activation and endorsing stereotypical beliefs by avoiding certain subjects; and (4) perceptions and anticipation of bias.

- **Perceived**—Research has shown that students' perceived racial inequality in school is negatively associated with academic outcomes [55]. Perceptions of inequality, discrimination, and/or stereotypes can affect career aspirations through the de-identification, devaluing, and disengagement associated with stereotype threat [54, 56, 57]. While this research focused on African American and Latino high school students, there is similar research noting that female high school students are more likely to perceive barriers to educational attainment than their male counterparts [60]. Additionally, stereotype threat has been shown to disenfranchise female students, inevitably leading them to consider changing majors to fields not dominated by males [57–59].

- **Stress and Coping**—Long-term exposure to discrimination, bias, stereotype threat, and inequity in educational access and opportunity contributes significantly to underrepresented student's perception of barriers to entry [61, 62]. This constant weathering effect can lead to greater resilience and striving harder to overcome obstacles [40], but more importantly, it can and does often lead to disengagement with a particular domain [54, 63]. Additionally, evidence suggests that high-achieving students who perceive these barriers and disengage with their given academic domain demonstrate lower performance regardless of aptitude [63, 64].

To validate their work, Scott and Martin conducted a study including quantitative and qualitative measures to determine the effect to which these barriers affected a cohort of students. They concluded that high-performing students of color perceived obstacles to success in STEM fields and this perception negatively impacted the likelihood of them pursuing STEM careers.

Of significant value is their finding that women of color face a double-bind, or cumulative effect of being both underrepresented with respect to race and gender.

It is critical moving forward that we understand these barriers to entry. Doing so allows us to better understand how to design programming, thus facilitating pathways and mentoring to contribute to a more inclusive, diverse experience for all students. We touch upon two specific approaches to addressing barriers in Chapters 5 and 6 where we focus on facilitating pathways and mentoring, respectively.

3.3   LEAKY PIPELINE

A long-held metaphor to describe the pathway from pre-school to Ph.D. or the pathway from pre-school to a STEM career, the STEM pipeline has long been controversial as a gross over-simplification of the complexities involved in recruiting and retaining underrepresented communities in STEM fields. A leaky pipeline refers to the unintended loss of people along the trainee path from their desired discipline. Contributing to the controversy, Garbee notes, "conversations concerning 'diversity' and 'equitable representation' in STEM fields began not as they should have—as a recognition of the value that comes from accepting and valuing a diverse set of perspectives and experiences in science writ large—but as concern about where to find all these extra people to make into STEM Ph.Ds. Suddenly women and minorities went from being widely disregarded by the scientific establishment to being seen as future economic contributors [65, 66]." The National Science Foundation has recently altered their nomenclature to refer to the pipeline as pathways, suggesting that there are multiple points of entry and multiple types of outcomes. Nonetheless, the fact remains that this "unintentional loss" is still very much a factor and one that must be understood and addressed.

Allen-Ramdial and Campbell proposed a new way of imagining the STEM pipeline as a vertical structure subject to the laws of physics. Downward forces, for example poor or non-existent mentorship, work against students moving upward in their educational pipeline [67]. They report the single largest rate of attrition in the educational pipeline occurs in undergraduate education, where students either leave the university environment or change to non-STEM majors to complete their degrees. They note that social identity, and a sense of belonging, are seen as powerful motives for long-term success [68], and suggest that the pipeline requires repairs and improvements through alternative or complementing strategies to support a student's identity as an emerging scientist. They note that increasing STEM participation requires greater efforts to retain those already in the pipeline and to change perceptions of the pipeline itself. Toward this effect, they suggest interventional strategies that act as opposing forces to known causes of attrition.

- **Align Institutional Culture and Climate**—The culture represents the collection of shared values and beliefs, while the climate represents practices and behaviors that determine prevailing attitudes. Research shows that the climate affects a student or trainee's sense of belonging [69]. While alignment of culture and climate can have positive effects on a student's participation, persistence, and long-term success [70], they often do not align and this misalignment can be detrimental [71, 72]. Climate surveys can be powerful means of capturing and monitoring issues that do effect climate [73, 74] providing data that can help inform and guide subsequent decisions.

- **Building Interinstitutional Partnerships**—One of the most powerful interinstitutional relationships is between faculty at majority serving institutions and faculty at minority serving institutions. This relationship, when truly collaborative, maximizes interaction between the institutions, affording greater access to underrepresented students that have participated in mutual social and academic activities, thus expanding the student's professional network and providing them with a greater sense of belonging and trust [75]. These types of partnerships are also instrumental in unmasking and addressing cultural differences and provide powerful means for developing cultural competency [76].

- **Attain and Sustain Critical Mass**—Achieving critical mass for underrepresented groups is one of the single most important indicators for a program's long-term success in achieving diversity and inclusivity. This is because it directly addresses one of the largest barriers that underrepresented students face—isolation [68, 77]. Feelings of isolation can inevitably lead to diminished sense of self-efficacy and enhanced feelings of stereotype threat [78].

- **Achieving, Rewarding, and Maximizing Faculty Involvement**—Faculty bring with them the ability to train and mentor budding students to become scientists in their respective disciplines. However, current institutional practices do not encourage or reward active participation by faculty. Faculty passionate in pursuing programs in diversity and inclusion often do so at the expense of what is rewarded—advancements in basic research. However, this does not have to be the case. Diversity and inclusion is an institutional commitment with many universities these days, recognizing that educating a diverse student population feeds directly into shaping the makeup of a diverse workforce [79]. Additionally, one of the primary factors for successful mentorship with all students, particularly underrepresented students, is persistent engagement. Otherwise, students often bounce from one well-meaning program to another without truly forming a network of mentors, reinforcing feelings of isolation.

3.4   COMMUNITY CULTURE

Improving outcomes for underrepresented students is highly dependant on the institutional culture and climate. As stated earlier, active and persistent engagement by faculty and administrators is key, but it is important to note that they are also stakeholders who should recognize themselves as beneficiaries of investments in diversity and inclusion efforts [67]. It is worth stating that the mere effort of trying to create a diverse and inclusive organization/team in and of itself contributes toward an inclusive community culture. What may be obvious by now is that this is neither simple nor is it to be considered short term. As stated by Lisa Keglovitz, Senior Vice President of Human Resources at GameStop, "Building a truly diverse and inclusive culture includes responding patiently and sensitively to the unconscious biases that currently exist [36]." Conversations and efforts in this space can be difficult, but nonetheless, they must happen. While we have no "silver bullet" to provide you in this space, we can in fact provide best practices for creating and maintaining a diverse and inclusive culture [80].

- **Establish a Sense of Belonging for Everyone**—Having a sense of belonging to something larger than yourself is not just a desire, it is a psychological need. That sense of belonging requires continuous efforts to ensure that team members feel as if they belong and that their opinions and voices matter.

- **Empathetic Leadership is Key**—Leadership with a personal connection or stake in diversity and inclusion has obvious benefits and can't be overstated as important to the overall culture. If leaders don't feel personally connected and passionate toward the mission, those in the organization will see that and consider it as less important.

- **A Top-Down Approach isn't Enough**—While top-down ensures compliance, it doesn't ensure commitment. An inclusive culture requires efforts that execute at all levels—top-down, bottom-up, and all points in between.

- **Quotas Don't Automate Inclusion**—While quotas may provide short-term gains in diversity, they don't speak to retention. Without an inclusive culture, retention will suffer and this will in turn affect abilities to recruit talented, diverse candidates.

- **Inclusion is Ongoing, Not One-off Training**—We all approach our lives with our own biases that we have developed over a lifetime of experiences. Recognizing these biases and opening our minds to other perspectives is key to creating a truly inclusive culture, one that is capable of having honest, tough conversations and affecting long-lasting change. This is an ongoing process that must be integrated into the fabric of the organization.

- **Maximize Joy and Connection, Minimize Fear**—Universities are hard-wired to operate in a fear-motivated environment. However, motivating through fear oftentimes causes students to shut down. Re-framing the conversation to speak from a point of possibilities creates greater potential for long-lasting change.

- **Forget Fit and Focus on Helping Individuals Thrive**—Fit can often be an exclusive factor that has legacy norms woven into it. A skeptical look at fit is well worth doing on a continuous basis to ensure that individuals capable of being valuable contributors won't be overlooked from the beginning. This necessarily means that we reconsider what it means to be smart and what expanded set of skills would indeed be valuable.

In summary, an inclusive and diverse culture requires persistent and ongoing attention to the many interconnected issues at play. It is not enough to simply be passionate about this;

rather this has to be a material part of all facets of the organization. Over the last few decades, the needle representing diversity in STEM, computer science specifically, hasn't changed. This is because we all do what we are trained to do when we see a successful program. We attempt to determine what contributes to the "magic" and subsequently attempt to reverse engineer and replicate as if it is simply a recipe that can be followed. History suggests that we will continue to fail to move the needle with this approach. Rather, building diverse and inclusive programs is hard and requires passion and persistence. While this may seem daunting on its face, the rewards of succeeding are undeniable.

CHAPTER 4

Case Studies

Contributors: Anonymous; Organized by Meg Pirrung

4.1 IMPORTANCE OF SHARING EXPERIENCES

"Before you criticize a man, walk a mile in his shoes"—proverb

The key to understanding someone and their challenges, is to build empathy for them. In this chapter, we seek to help the reader understand what it is like to be in an environment that is not inclusive. This is particularly difficult for someone who does not regularly deal with the direct and indirect effects of environments that are not welcoming to underrepresented groups. Rather than attempt to explain, we instead simply recount stories from members of the visualization community. These anonymous stories are real. They all involve people from our community. It is important to note that these case studies represent a small fraction of the individuals we spoke with. Many of them chose not to come forward with their story for fear that they may be identified and would suffer retribution.

4.2 DESCRIPTION AND FRAMING OF CASE STUDIES

Each case study in this chapter consists of four parts; Setup, Situation, Outcome, and Takeaway. **Setup** helps contextualize the case study for the reader. Any relevant information including background of the individuals involved, setting for the encounter, and prior history will be included here. The **Situation** section describes the actual event that took place in which the affected person was injured[1] by the offending party. Situations may be one-off instances or repeated attacks of varying severity. As actions or statements of high severity may have more obvious next steps or actions, it is important to also illuminate instances of injury or micro-aggression that have unseen effects and outcomes. We relay these situations in first person in an attempt to encourage empathy in the reader. Our goal is to ultimately assist the reader in both recognizing and understanding how certain actions and statements affect others. **Outcomes** have four different categories: (1) direct for the affected individual(s); (2) direct for the offending individual(s); (3) direct for bystanders; and (4) institutional. Categories 2 and 3 can be especially

---

[1]In this context, we define *injury* as any statement or action said or performed either directly or indirectly to an individual or group of individuals who are made to feel unsafe, unwelcome, or unduly burdened. An example of undue burden which is often overlooked by those who are not part of a marginalized group includes not feeling like you can bring your whole self to a situation, e.g., a black woman may police her own tone and speech for fear of being labeled as an "angry black woman."

difficult to ascertain as these case studies are relayed by the affected parties. Effective conversations on the topic of diversity and inclusion will contribute to positive outcomes for all parties. The **Takeaway** section serves to highlight specific elements of the situation and outcome that work to instruct the reader in recognizing problematic situations. Recognizing that situations are problematic is the first step to ideally prevent, interrupt, and remedy such situations.

4.3   CASE STUDY 1

Setup

IEEE VIS conference in 2017 in Baltimore, MD. The affected party was attending IEEE VIS for the fifth time.

Situation

I entered an elevator in the conference hotel to head down to find lunch. As I enter, I observe two other conference attendees—the lanyards and name badges a dead giveaway. They too were heading down. I punched my number and did the socially acceptable thing in an elevator, faced forward and stared at the door. The other two attendees began speaking to one another. I noted the foreign accent. I wonder if they are international attendees? From the accents, I assume an Asian country.

IA1: Where do you want to eat?

IA2: I don't know, do you have any ideas?

IA1: I'm not sure. But I don't want to go outside the hotel.

IA2: (silence)

IA1: It's not safe. There are a lot of black people out there.

IA2: (silence)

[Did I mention, **I am a black man**]

I stand there, stunned, in silence. I think to myself, "did he really just say that?" I'm angry, I'm hurt, I'm disgusted. Do they see me? Do they understand how insensitive, offensive, hurtful, and racist that is? Is this a cultural thing? The questions swirl. The elevator stops. The bell dings. I think to myself and can only mutter "Unbelievable." I exit the elevator without looking back.

I have lunch and spend the next hour replaying the experience in my mind. I can't focus on anything. I'm frustrated. I'm angry. Not only at them, but at myself. I should have said something. I should have "educated" them. I should have expressed how inappropriate that was. To lessen the pain and frustration, I confide in several friends at the conference. Telling my story makes it a little easier to deal with.

I was frustrated and angry with this person. But more importantly, I was angry with myself. How fair is that? I did nothing wrong. I was wronged...but still, I was angry with myself.

Outcome

In this situation there are several outcomes of note. The subject is made to feel unseen and invisible, contributing to feelings of isolation in the community. Second, speaking with friends about the situation is helpful. Third, there is no observable outcome for the [offenders], as they potentially did not even realize the statements were hurtful.

Takeaways

There are two important takeaways from this story: (1) the continued impact on the marginalized person and (2) a question of responsibility.

First, the subject of this encounter states that they were angry with themselves. Despite the fact that they are the victim of a [racist remark], they are angry with themselves. This is quite common for many underrepresented minorities [172]. Failure to respond appropriately in the moment leads to disappointment in oneself in addition to the feelings from the encounter. Unspoken and invalidated feelings may lead to hours, days, weeks of reflection and playback. These types of encounters are ubiquitous for people of marginalized groups, and these feelings compound throughout their careers and lives.

Perhaps this is a cultural issue. Perhaps the offender is from a country, a city, a town, where they experience few, if any, people of color. Perhaps they are simply ignorant. Who bears the responsibility to educate them so that they do not offend others at the conference?

4.4   CASE STUDY 2

Setup

This happened at a high-profile conference in our series of related conferences. The affected party had just entered the Visualization community as a student.

Situation

I was a student that had just recently begun attending IEEE VIS and related sister conferences. I was eager to learn and confident that visualization would be my chosen field for many years to come. I was attending the conference with my advisor in a city I had never been to before. My advisor asked if we could work on some research and since I had the larger hotel room, he suggested we work in my room. Typing those words to this day still makes me feel stupid, as if as a young student I should have foreseen what would happen next.

I agreed that we could meet in my room, and later that day my advisor showed up. We discussed anything but research and I found myself playing a game of "keep away," realizing far too late that my expectations were vastly different from his. I replayed every off-color remark he had made at other times and asked myself why I didn't see this coming. He propositioned and I carefully declined. He suggested and I ignored. He made comments and I changed the subject. All the while, the dialogue in my head was rooted in confusion. I kept replaying every response, every reaction, and wondered what **I** did to land me here. I was young and didn't have the benefit of foresight, of wisdom to know that the only thing I was guilty of was naiveté.

I managed to get him out of my room after some careful coaxing, but the really damaging piece of this happened long after. I could not bring myself to leave my hotel room after that until I left for my flight home. I had a hard time getting out of bed. Rather, I kept replaying everything that was said, forensically analyzing all my actions and responses. I felt ashamed and stupid. My partner called me and he must have heard it in my voice. I told him everything, not knowing that he would struggle to process as well. He went and did what he believed was in my best interest. He reported to someone higher up, to my advisor's boss.

When I returned home, I was called in to my advisor's boss' office. I don't remember much from that time, but I remember this as if it happened yesterday. Rather than asking me what had happened, he began by telling me that my advisor had a family, one that needed his financial support. He told me the impact it would have on his life if I decided to move forward and then he said, "This is a serious allegation that has been made. This didn't really happen, did it?" I knew at that moment, if I wanted to continue along in my career, I had only one choice. I did not recant what had been told him. Rather, I thanked him for his time and got up and left his office. I was furious, confused, and embarrassed. In many ways, that moment was a turning point for me. All the enthusiasm I had previously had with Vis and all the possibilities I had felt prior to that incident were wiped away.

Outcome

Significant time has passed since that incident, and today's culture is on its face very different. I chose to speak up again when I realized that young women that are now my age when this happened to me are continuing to struggle with these issues. That realization was like a blow to the head. That incident, and so many others like it, have indelibly shaped my career, and the way I maneuver in a world where I am often the only woman in the room.

Takeaways

Instances like this have long-lasting repercussions for those of us that have experienced them first-hand. They shape the way we interact with our male colleagues. They shape our career decisions, and often-times they determine whether or not we stay, knowing that the odds are good that something like that will happen again. As a community, we speak of our male colleagues in terms of their intellect and our female colleagues in terms of their appearance and their personality, not realizing that by doing so, we feed into this stereotypical pattern that so often lands some of us in these raw, inappropriate, and damaging situations that take a lifetime to overcome.

4.5   CASE STUDY 3

Setup

This happened at the IEEE VIS conference. The affected party was a student giving their first IEEE VIS conference presentation.

Situation

A couple of years ago I was at IEEE VIS to give my first paper presentation. I had been to the conference once before so I was aware of what a big deal and honor it was to present. I practiced my talk over and over again receiving invaluable feedback from my Ph.D. advisor and peers. I practiced at home, in the lab, while walking, and even in front of my bathroom mirror. I love fashion, and more importantly am a "dress for success" kind of girl—a new outfit always makes me feel special and confident, and for my first IEEE VIS talk I needed all the confidence I could muster! Thus, I made a weekend shopping trip out of it, and found the absolute perfect black pencil skirt, black and grey ruffle edged blouse, and smooth black leather flats. The ensemble was not too fancy, nor too casual, it was just right.

I practiced and practiced, got up on stage absolutely terrified, but felt good about myself in part because of my outfit, took a deep breath, and knocked the presentation out of the park! I nailed it, better than any of my practice talks. I knew it from the energy of the room, the hum among people, and the complimentary comments given along with the questions during the Q&A. After the end of my talk session I answered a couple questions at the front of the room and went out for the coffee break. The first person I saw walking toward me was a famous senior female member of the IEEE VIS community, someone I idolized and cited in my just-presented paper. I started gasping for breath realizing she was coming to talk to lowly me! However, before I could even try to fumble through saying "It's an honor to meet you!," the women looked me straight in the eye and said "Good talk, but lose the dress. No one will ever take you seriously in a dress, and don't act so cutesy."

I was absolutely stunned, and felt like someone had kicked me in the stomach. I honestly do not recall what I said in reply as I was in such a state of shock, but it was something along the lines of "Oh, <silence>, great to meet you." and then the woman walked away. Instead of reveling in my success, and soaking in all the compliments from my fellow attendees, I instead went up to my hotel room and cried. Going into and immediately after the talk I felt so good about myself, and loved the way I looked, never questioned my appearance or what effect it or my gender could have on my credibility, but apparently that is not how I am supposed to dress for success in the IEEE VIS community.

Outcome

It took a couple weeks for me to bounce back, but I am blessed to have a very supportive family and peer group and everyone gave me the same message: I need to be me, and if the community cannot accept me for me then it is the wrong community for me to be a part of. Although I have felt intimidated and uncomfortable seeing the senior women at IEEE VIS since our encounter, I have continued to wear my confidence-boosting fashionable outfits at IEEE VIS.

Takeaways

Appearance is often the first thing we notice when we interact with someone and, unfortunately, comments like this happen far too often to women. While the comment was likely intended to be constructive, it perpetuates an emphasis on appearance rather than content. These kinds of seemingly harmless comments very often have long-lasting effects and contribute to a negative culture by reinforcing stereotypes. While we often think that men are the only ones to make comments focusing on appearance, it is important to note that this comment was made by a woman.

4.6   CASE STUDY 4

Setup

Junior member of the community. First major public speaking opportunity.

Situation

After the session in which I presented was over, a few people from the audience gathered to thank me for my talk and ask questions. My session is the last of the day, the next event is the conference dinner. Various attendees are asking questions, and it is getting close to time for the dinner.

One attendee steps forward and states that he has to go meet some of his friends for the dinner, but that he'd like to shake my hand and thank me for my talk. After shaking my hand he says "You have such a strong handshake for a beautiful woman." I am taken completely off-guard and can't think of anything to say. I think my feeling showed on my face, and after this person walked away a couple of the attendees in the circle laughed awkwardly. The man walks away without noticing my dismay.

My science should be centered in this context, so why is my appearance brought up, regardless of the handshake? How does a woman's appearance have anything to do with how she shakes hands?

I wish I would have said something along the lines of "What is that even supposed to mean?" or even just an incredulous "What?" I was genuinely baffled at the statement.

Unfortunately, I have been in similar situations where a man intends to "compliment" a woman, and instead offends. When I have spoken up in these situations, the response is often along the lines of "Can't you take a compliment?" and "Why would you be offended by that?" This sometimes makes me think, "Why *am* I offended by this?" I think that the reason is the statement insinuates that as a described "beautiful woman" that I shouldn't have a strong handshake, and that in having one I am somehow strange.

Outcome

Again, the offended individual bears the brunt of the incident—conveying disbelief, feeling anger and feeling guilt for not having done anything about it. The offending individual shoulders none of it. In this particular case, others were witness and involved, and it appears felt uncomfortable with the situation as well. Perhaps lasting effects were left with them as well. How will they view the offended individual? How will they view the offender?

Takeaways

These incidents effect all involved. For the offended, in particular, it's often difficult to respond in the moment. For the bystanders in the circle, this was an opportunity to make a difference. Perhaps they were shocked and offended as well. However, in the interest of nurturing an inclusive environment, we all bear a responsibility to speak up or at least acknowledge the awkwardness caused by the offender.

4.7   CASE STUDY 5

Setup

This happened to a female member of the VIS community as she was coming up in the field.

Situation

I am a member of the IEEE VIS community and when I was coming up in the field, I had an issue with a powerful male member of the community. Nothing was quite so horrible that I tried to find a way to report it (and frankly at the time I would have had no clue who to report it to), but it made me uncomfortable enough that I avoided this particular individual as best as I could (but I couldn't always do so). Honestly, for the most part, it was the kind of thing that I could ignore and move on. I was uncomfortable but not intimidated, and it was a different time for women in the field. Many of the details of my story are pretty vague (male colleague makes me uncomfortable by sitting too close and talking to me too often), and perhaps this could be attributed to a cultural difference rather than intentional harassment. However, this individual also liked to tell inappropriate jokes. I do vividly remember my breaking point with this male colleague in our community—a particular "joke" that I found so offensive that I vowed to myself at that point that I would simply not have anything more to do with him (even if it meant excusing myself immediately from every group conversation he joined).

The joke was "Why is it better to have sex with 20 two-year-olds rather than 20-year-olds?" and after a shocked silence where no one in the small group was willing to offer a guess he said "They may have less experience but hey, there's 20 of them."

For me, I know it was very hard at the time to think of the right thing to say in the moment, especially given that the joke was said to a small group of folks (with everyone but me being male). Now, I am better at dealing with these things by taking myself out of the equation and instead acting as a champion for others—but then (and sometimes even still now) it was difficult to form the right words on the spot to say in front of a group of folks. For me at least, in situations like this, my brain goes into overdrive trying to think through every question from, "What did I do to indicate I'd be OK with this behavior?" to "What are the consequences of my saying something?"—which gets harder the larger the group you have to say it in front of.

While this was an isolated situation for me at VIS (of blatant inappropriate behavior), it has not been an isolated incident in my professional life. There was a professor in graduate school who liked to comment on my appearance in inappropriate ways, often standing close enough to me to look down my shirt and make a comment on my bra, or tell me that something I was wearing looked particularly slutty. There was a fellow student in a large physics lecture who would try to sit next to me so that he could say suggestive things and then rub up against me as we filed out of the auditorium. There was a manager at one of my previous jobs that never looked at me, even if I asked a question in a meeting (he'd almost always phrase his response to one of my male colleagues, as if he asked the question instead of me).

Outcome

Now, I primarily see sources of unintended bias rather than blatant harassment and discrimination, and I believe we are making progress. I wonder if I don't see the blatant issues as an older woman that I would still be seeing if I were in my 20s or 30s. I now understand how damaging this sort of behavior can be to the community as a whole, and understand the importance of everyone speaking up when this type of behavior happens.

Takeaways

Sexualized jokes are not appropriate in the workplace and this extends to any professional setting. They can be graphic and shocking and they can reinforce negative stereotypes, making others feel self-conscious and uncomfortable. These types of jokes are the most common form of harassment in academia and contribute to making women feel excluded and isolated. These types of instances contribute to an environment that makes it difficult for women to excel. Women, particularly in male-dominated fields, very often feel as if they are powerless to speak up. Male allies in the field who actively speak up are key for successful change in male-dominated fields. The National Center for Women and Information Technology (NCWIT) provides strategies male allies can use to accelerate positive change [21].

CHAPTER 5

Community On-Ramps

Vetria Byrd and Kelly Gaither

Visualization is a catalyst for communication, a conduit for collaboration, a pathway to STEM, and a mechanism for broadening participation [106]. This chapter presents visualization initiatives and strategies to broaden participation of members of underrepresented groups in the U.S. For the context of this chapter underrepresentation extends beyond gender, race, and ethnicity; it includes geographic location, lack of access to resources that facilitate engagement and inclusion in the field, and persons with disabilities in the U.S. Section 5.1 describes recruitment and onramp efforts to broaden participation in visualization through a series of biennial Broadening Participation in Visualization (BPViz[1]) workshops open to all persons with an interest in data visualization. Section 5.2 describes an onramp option to engage undergraduates in data visualization while Section 5.3 describes onramp efforts that engage undergraduates in data visualization by teaching concepts in the context of socially relevant problems.

5.1 BROADENING PARTICIPATION IN VISUALIZATION (BPVIZ) WORKSHOP

Visualization plays a significant role in the exploration and understanding of data across all disciplines with a universal goal of gaining insight into the complex relationships that exist within data. In 2014, the first Broadening Participation in Visualization (BPViz) Workshop was hosted at Clemson University in Clemson, South Carolina. It was the first discipline specific workshop co-sponsored by Computing Research Association (CRA) and The National Science Foundation (Award #1419415) to focus entirely on the introduction of data visualization to groups underrepresented in the field. The success and the level of interest in the first BPViz workshop [107] led to the biennial BPViz Workshop series. BPViz'16 and BPViz'18, hosted in July 2016 and June 2018, respectively, were co-located at Purdue University and the University of Illinois Urbana-Champaign. The primary objective of the workshop is to broaden participation of women and underrepresented groups in visualization.

---

[1]http://bpviz.org/

5.1.1 SUPPORTED ACTIVITIES

The organization of BPViz is intentional and aims for diversity in speakers and participants. Respected and leading members of the visualization community, particularly those who are members of underrepresented groups, from academia, research, and industry are invited to participate in the workshop as speakers and/or panelists. Participants are generally new to data visualization, have limited or no experience with data visualization tools, and are unaware that data visualization is a process. BPViz provides an introduction to the data visualization process, visualization applications, and hands-on experience with current visualization tools.

There are many pathways to visualization and BPViz provides the essential initial connection between workshop participants and industry experts. The eclectic journeys of persons in the field are shared with participants as part of a Pathway to Visualization Panel. Their stories help participants envision themselves in the field, see that they have many career options, and think about how to incorporate data visualization in their daily practice. A signature feature of the BPViz Workshop is the "Meet the Panelists" session. Participants are given an opportunity to talk with each panelist in a smaller group setting where they can ask additional questions or engage in an extended discussion of a panel topic with the panelist.

The workshop is designed to inform, inspire, and encourage participants to engage in the multidisciplinary dynamics of visualization. One of the expected outcomes is that participants will become a part of the larger visualization community. To encourage participation, and to give participants a feel for what to expect at a visualization conference, the workshop mimics, on a smaller scale, some features of the IEEE VIS Conference. Specifically, participants are asked to create a poster for the poster session and to give a one-minute slide presentation to describe the highlights of their work and to encourage fellow participants to visit the poster for more scholarly dialog.

5.1.2 PARTICIPANT RECRUITMENT

Participants for the workshop are recruited via a number of different venues including, but not limited to: the Anita Borge Institute/Grace Hopper listserv, association listservs (CRA, CRA-W, CDC, AAPHDCS, Hispanics in Computing, and Women in Machine Learning), Broadening Participation Communities (A4RC, EL Alliance), and discipline-specific mailing lists (IEEE-VIS).

Other potential avenues for fostering community onramps might include reaching out to specific demographics of interest: historically Black Colleges and Universities (HBCUs), Hispanic Serving Institutions (HSIs), Minority Serving Institutions (MSIs), and Tribal and Community Colleges. The first year of any broadening participation or outreach effort requires an investment in time and effort to reach the desired target audience. For BPViz, the call for participation is intended to reach all persons with an interest in data visualization. Past participants of the workshop have included: K-12 students, undergraduates, graduate students, post-docs, junior faculty, senior faculty, research scientists, instructors, lecturers, administrators, artists, and

IT and business professionals. BPViz provides a high-level introduction to data visualization in an intense, but fruitful, two-day period.

5.2   VISREU SITE: COLLABORATIVE DATA VISUALIZATION APPLICATIONS

Studies have shown that the exposure to current research early in the careers of all students encourages them to continue on in the pursuit of advanced degrees [108]. This section describes the on-ramp efforts to broaden participation in visualization at the undergraduate level through research experiences for undergraduates.

Data visualization plays an important role in all levels of scholarship. Visualization is the process of transforming raw, complex data into a visual representation that provides insight. In order to prepare the next generation of researchers and scientists to make transformative and innovative discoveries in a data-driven world, exposure to the process, tools and techniques of data visualization must begin early [109]. In 2014, the NSF REU Site: research Experiences for Undergraduates in Collaborative Data Visualization Applications (Award #1359223, Vetria Byrd, PI) was the first Research Experience for Undergraduates (REU) site with a primary focus on introducing undergraduates to the data visualization process. The eight-week program, Vis-REU, was designed to: (1) introduce data visualization at the undergraduate level; (2) strengthen student skills and capabilities in data visualization; (3) broaden participation in visualization among women, members of underrepresented groups and students from institutions with limited research infrastructure; and (4) encourage students to pursue graduate degrees in STEM. Visualization training was part of the core summer curriculum.

Undergraduates from both STEM and non-STEM majors participated in the program [109]. Much like the traditional research experience model, students were paired with research faculty and joined a research team with specific research tasks and duties. What was unique about the VisREU site is that students and their faculty mentor were paired with a data visualization mentor; someone knowledgeable of the data visualization process with experience transforming data into insight. The visualization mentor worked with the student and faculty to ensure the desired outcome when visually representing the data. The program included multidisciplinary research projects in both the traditional STEM disciplines (computer science, engineering, genetics and biochemistry, sociology, molecular modeling and simulation, and inorganic chemistry) as well as typically unrelated disciplines such as athletics. Students participating in the summer program reported majoring in engineering, computational biology, computer science, engineering, mathematics, and information systems.

BPViz workshops are open to all persons, regardless of academic status, with an interest in data visualization. The VisREU site targeted a specific demographic: underrepresented undergraduate students (as defined above). The recruitment effort mirrored the recruitment effort for the BPViz Workshop with marketing to listservs, broadening participation communities and discipline specific mailing lists. The ideal approach to recruitment is to establish a relationship with the groups of interest prior to recruitment. In the absence of an existing relationship, the first year of recruitment for the program was methodical and time intensive but yielded favorable results. The most effective strategy is simply to identify your audience and then reach out to them directly.

- In 2014 The Obama Administration released the White House Initiative on Historically Black Colleges and Universities School Directory [110]. The document contains web addresses for each school. Every website in the directory was visited to identify a contact person at the school's career services office. Once identified an email was sent to the attention of the Career Services Office with details of the summer research opportunity.

- A similar approach was taken to identify Hispanic Speaking Institutions (HSIs). A Google search returned the website for the Hispanic Association of Colleges and Universities [111]. This website provides a list of HACU Member Hispanic-Serving Intuitions, by state and a HACU-Member HIS Map. As of this writing HACU has 298 HIS Members in 20 states and Puerto Rico. In 2014, all of the member institutions with a website with information on who to contact at their career services department was sent a recruitment email.

- Community Colleges were also considered to be underrepresented and therefore target recruiting grounds. Due to the vast number of community colleges in each state, it might be advantageous to identify which states to target and develop some criteria or metric for choosing which colleges to reach out to.

In instances where campus career services were not found on a website as identified above, a faculty member or instructor whose campus online course information aligned with the goals of the VisREU site, or who showed some area or domain of interest that could benefit from data visualization, was contacted via email. The call for participation email also included a statement indicating all undergraduates, especially those from underrepresented groups, who had an interest in data visualization were encouraged to apply. This initial process is tedious and time consuming, yet fruitful. It assumes up-to-date web pages with correct contact information.

In 2014, the program received 26 applications and in 2015, 205 applications for 10 slots. The VisREU site hosted a total of 22 participants during the 2014/2015 program: 11 (50%) female, 11 (50%) male; 4 (18%) participants were from historically black colleges or universities (HBCUs), 1 (5%) from Hispanic serving institutions (HSIs), and 5 (23%) were first-generation college students. The program provided diversity in content, projects and participant ethnicities: 6 (27%) African American, 1 (5%) Asian, 4 (18%) Hispanic/Latino, 1 (5%) Native American/American Indian, and 10 (45%) of the 2014/2015 cohort were from non-Ph.D. granting institutions. The combined 2014/2015 cohorts reported: 21 accepted student conference poster presentations, 15 accepted student conference talks, and 55 student REU site presentations (including midterm, final presentations, and presentations to incoming freshmen from underrepresented groups) about the summer program and their research.

The VisREU site was timely. It took place at a time when there was renewed interest in visualization, there were questions about how to make sense of large amounts of data (i.e., big data), and visual analytics was beginning to play an important role in addressing these questions. The increase in applicants for the summer program suggests there is growing interest in data visualization at the undergraduate level, across disciplines and among STEM and non-STEM majors.

5.3   ADVANCED COMPUTING FOR SOCIAL CHANGE AND COMPUTING4CHANGE

Over 20 million students are enrolled in our nation's post-secondary education system of 2-year through Ph.D. and professional degree granting institutions. Forty-two percent of these students are African American or Black, Hispanic or Latinx, Native American, Pacific Islander, or multiracial. Yet, the profile of enrollment and graduation in STEM fields does not mirror the general population of the U.S. or their enrollment in post-secondary education. Only 19% of students enrolled in engineering are female, 8% are Hispanic or Latinx, and 5% are African American or Black. Less than 3% of the students pursuing degrees in computer science are Hispanic or Latinx, the fastest growing demographic in the U.S. accounting for more than half of the total U.S. population growth from 2000–2014.

We designed Advanced Computing for Social Change (ACSC) to test out our theory that teaching complex science and technology skills in the context of larger socially relevant problems of interest is an effective way to educate all students and a particularly effective way to engage underrepresented communities, lending credence to our belief that it is crucial to teach students "how" (the mechanics)in the context of "why" (the motivation). Engaging these diverse communities in ACSC is an effective way to accelerate scientific discovery and contribute to the national economic prosperity as called for in the Executive Order—Creating a National Strategic Computing Initiative [112]. The primary goal of the ACSC is to actively involve students in solving problems that may seem beyond their knowledge scope, but are relevant to them personally, and ask them to produce tangible outcomes. This teaches students to not only be critical thinkers, but to be creative and reflective as well. The ACSC curriculum is built on four primary tenets [4].

- *Storytelling:* Neuroscience is discovering that the brain is wired to organize, retain, and access information through story [113–115]. Teaching through story aids memory, provides building blocks for learning, and encourages imagination [116, 117]. Storytelling also puts information into an emotional context and emotions play an essential role in both memory and motivation.

- *Visualization:* As technology advances, we find ourselves in an increasingly digital world, oftentimes inundated with data. This growing mountain of data requires visual intervention or intermediate analytics to make sense, given inherent inefficiencies of our brain to process raw data. Brain research tells us that we will only remember 10% of information that we hear, but if we add a picture, we will remember 65% [118–120].

- *Team Science:* Dictated by the need to solve larger and more complex problems, there is increasing need for students capable of working productively in multidisciplinary teams. Research suggests that multidisciplinary teams are the most successful in fostering innovation [121]. A curriculum that actively fosters collaboration has proven successful at increasing productivity, retention, and success of women and underrepresented minority scientists in multiple STEM fields [122, 123].

- *Discovery-based Learning:* Active learning allows students to develop and implement planned activities as a partner in the activity. Evidence suggests that students learn and retain more by doing and being actively engaged. Active learning increases mastery of technical skills, improves communication skills, and enhances critical thinking [124–126].

In the initial ACSC, external evaluators conducted a pre-event survey, in-person focus group sessions, and a post-event survey. All students responded to the pre-event survey; 95% of the students participated in the focus group session, and 86% of the students responded to the postevent survey. Overall, the students rated the experience as 4.5/5.

The program theme empowered students to reflect on and intentionally influence social change utilizing their STEM background knowledge and skills. Prior to participation in the ACSC challenge, students reported a perceived lack of ability to affect social change due to a misalignment with their technical STEM backgrounds. Pre-event survey respondents commented on their desire to learn more about the connection between computing and social change. Specifically, students wanted to familiarize themselves with existing work and learn how to formulate ideas that lead to social change. Of the pre-event survey matrix items designed to gauge student interest and expectations, respondents rated Q1H "I expect to gain knowledge/skills in this program that will help me affect social change," very highly at 4.53 (Mean = 4.53, SD = 0.94, N = 19) on a scale of 1 (strongly disagree)—5 (strongly agree). Post-event survey respondents felt they achieved this expectation indicated by their high rating of item Q1L "I gained knowledge/skills in this program that will help me affect social change" (Mean = 4.50, SD = 0.63, N = 16; scale 1, strongly disagree—5, strongly agree). During post-event onsite focus groups, students expressed their newfound confidence to tackle social issues using advanced computing. They describe themselves as novel "sweapons" for moving social agendas forward.

Team-building activities facilitated later conflict resolution and appreciation for the team-based science approach to addressing social change. Students described the Welcome and Introductions dinner event and corresponding "Dare Greatly" activity as humanizing. Some expressed feeling initially "overwhelmed" and that they "couldn't compete" at the conference. The activity aided in easing some of their concerns through personal storytelling and creating a connection between participants. Unexpected conflicts arose during execution of ACSC challenge activities within teams. Some students were initially surprised by the additional task of navigating group dynamics, however, they described successfully "coming together" despite their differences. Post-survey respondents rated items regarding engaging with other students and plans for keeping in contact highly. On the same survey, students were asked to rate whether their own background knowledge/skills were sufficient to complete the challenge as well as their group's skills as a whole (Q1F and Q1G). Ratings (scale 1, strongly disagree—5, strongly agree) for their individual ability were lower (Mean = 3.73, SD = 0.96, N = 15) than ratings for their overall group's knowledge/skills (Mean = 3.87, SD = 0.92, N = 15) potentially indicating the value students found in working as a team to address a complex social issue. Respondents are also interested in serving as mentors in future iterations of the program.

ACSC created a safe space within advanced computing for students, particularly minority women, who regularly experience discrimination in the scientific community. Students, especially women, gave detailed accounts of discrimination they commonly experience while pursuing a career in science. Stories demonstrate feelings of occasional inadequacy and frequent frustration. Despite these, at times, alarming accounts, students expressed their gratitude for the environment where they can "just be people." Over two-thirds (64%) of students cite a lack of training and support at their local institutions and would like access to materials both before and after the challenge. Of the 14 students who commented on the pre-event survey Q9 "What challenges do you experience in your current program as you seek to gain expertise in advanced computing?", 64% reported that a lack of training, materials, courses, and support were their largest challenges. Regarding pre-event survey Q8 what topics they would like to learn, 5/15 commented on wanting to learn more about visualization. Almost half (47%) agreed or strongly agreed with the statement "I applied to this program because my institution does not offer courses in this area" and 79% expected to apply the knowledge/skills gained in this program to their work/research. During focus groups, some participants felt their background in coding was insufficient for participation the challenge. Many would like more training resources both before and after the program. Following the event, 67% (10/15) reported having sufficient background knowledge and skills to complete the challenge. Nearly all postevent survey respondents (94%, 15/16) plan on applying for the SC17 student program to gain further skills and contacts, as suggested by focus group comments.

While students enjoyed the program, many felt it did not reflect their expectations and made suggestions for further improvement. Focus group participants believed the challenge would mirror past student program events. Many were unclear regarding the platform or tools needed to participate in the challenge, what research questions to prepare for, data sources to investigate, and program leadership's expectations for final product quality. Pre-event survey respondents expected the program to be difficult (Mean = 2.67, SD = 0.67, N = 18; scale 1, very difficult—5 very easy), however, post-event survey respondents rated their satisfaction with the delivery format of the program highly (Mean = 4.27, SD = 0.74, N = 15; scale 1, strongly disagree—5, strongly agree).

A subsequent version of the program occurred in 2017 co-located with SC17 in Denver, CO. Ten students from the local Denver area participated and the topic was immigration. As a result of the success in 2017, the Association for Computing Machinery Special Interest Group on High Performance Computing (SIGHPC) picked up sponsorship of the challenge, renaming it to Computing4Change [127] and making it a competition. The inaugural Computing4Change competition was held at SC18 in Dallas, TX where students learned data analysis and visualization methods while exploring issues related to violence. When applications opened in Spring 2017, SIGHPC received hundreds of applications from citizens of 31 countries, with over 60% coming from non-U.S. citizens. Forty-five percent of applicants were female or non-binary gender, and 4% of applicants identified as having some form of disability. Of the applicants from the U.S., over half were from groups underrepresented in computing. More than 160 universities were represented, including students from more than 100 majors ranging from law and urban planning to engineering and economics.

The applications were reviewed and evaluated by a panel of experts from diverse backgrounds across race, gender, discipline, and nationality. Selections were based on applicant's vision for using technology to affect positive change in an issue relevant to them; overall potential for impact in their chosen fields and home institutions; and the extent to which they can serve as ambassadors to increase diversity in the workplace.

The 16 chosen participants were citizens of 5 countries. Sixty-three percent of awardees identified as female, 12% of awardees identified as having a disability, and 50% had never attended a professional conference. Among awardees from the U.S., 33% were Black/African American, 25% Latinx, 19% White, and the remaining were Asian, Native Hawaiian/Pacific Islander, or of Mediterranean descent. Students were from 16 different universities, 44% of which are classified as resource constrained by the Carnegie Classification of institutions of higher learning. Competition details were revealed onsite at SC18. Topics for the competition were crowd-sourced from applicants and included potential topics such as gun-control, climate change, education, racism, food security, clean water, and sexual harassment.

CHAPTER 6

Retention

Ron Metoyer, Manuel Pérez Quiñones, Anastasia Bezerianos, and Jonathan Woodring

In the previous chapter, Drs. Vetria Byrd and Kelly Gaither presented several successful models for engaging a diverse population with the visualization community through effective onramp programs. These focused, deliberate efforts are necessary to truly engage and include a population that is not already well represented within the greater research community. In this chapter we focus on retention—that is, what should a community do once a diverse population engages in order to keep them engaged and more importantly, help them thrive in the community.

6.1 MENTORING AND COMMUNITY BUILDING—CRITICAL TOOLS OF RETENTION

According to the National Academies of Science, efforts to recruit and retain underrepresented minorities must be urgent, and sustained; comprehensive in addressing the full pathway pipeline; intensive to address inadequate social, educational, and financial support; coordinated across groups and organization; and informed by best practices [128]. An *intensive* effort, in particular, is a focused intervention that seeks to fill in the gaps and level the field for those who have not had the same level of exposure to STEM as others. These gaps may be in financial support, mentoring, social integration, and professional development. Similar recommendations for focused interventions have been made with regards to retaining women in STEM [129]. In this section, we will examine targeted efforts to mentor and integrate underrepresented groups into a research community.

Mentoring is a crucial component of the success of graduate students, post docs, and junior faculty alike, particularly for women and underrepresented minorities. Mentors serve as role models, supporters, advocates, advisers, and more. While the "match" model of assigning a more senior experienced person (e.g., Full Professor) to a less experienced mentee (e.g., Assistant Professor) has historically been employed in many institutions and organizations, a new "network model" has emerged as a popular alternative in recent years [130]. In this model, (1) the individual mentor is replaced or augmented by a network of mentors where each fills a specific mentoring need for the mentee and (2) the mentee is an active participant in creating that mentor network. While the traditional model has merits, some argue that this network model (also known as multiple mentoring) is even more effective in providing multiple targeted sources of accountability as well as resources for collaboration and sponsorship in various aspects of the academic career. Implementing a mentoring policy such as this is typically not the responsibility of a research community, but rather an employer. However, supporting the building and utilization of such networks can be accomplished by a determined research community and can benefit that community in the long run. In the next section, we present two efforts, CHIMe and VisBuddies, from two different conferences, that attempt to create the environment in which these kinds of networks can be formed and cultivated in support of graduate students in one case, and in support of newcomers to a research conference in the other.

6.2   THE CHI MENTORING (CHIMe) WORKSHOP

The Coalition to Diversify Computing (CDC) is a joint organization of the ACM, CRA, and IEEE-CS that was founded in 1996 by Dr. Sandra Johnson and Dr. Andy Bernat with the mission of aiding in the building of a diverse community of computing researchers and professionals. The CDC spearheaded several major projects from 1996–2016 (when it was dissolved) including the organization of the Richard Tapia Celebration of Diversity in Computing Conference,[1] which still runs today. The CDC also successfully partnered with the CRA to offer many successful discipline-specific mentoring workshops[2] designed to provide mentoring for graduate students in specific research communities.

In 2010, Manuel Pérez Quiññones and Ron Metoyer organized the first discipline-specific mentoring workshop for the Human-Computer Interaction (HCI) community, co-located with the SIGCHI (CHI) conference in Atlanta, GA. This initial workshop, CHIMe, was financially sponsored by the Empowering Leadership Alliance, Virginia Tech University, the CDC, and the CRA. The workshop was designed to bring together, in one place, a unique, talented group of underrepresented students doing research in the human-computer interaction field to:

- help the students build relationships/networks with their peers;

- help students build relationships/networks with leaders in their field;

- provide a welcoming environment for mentoring and collaboration; and

- encourage the students' participation in the leading research venue in their field.

In short, the goal of CHIMe was to help students establish the building blocks of a mentoring network and to support their integration into the broader CHI research community. It provided a unique forum that both highlighted the technical achievements of diverse professionals but also facilitated the mentoring, networking, and honest feedback that is invaluable to graduate students.

[1] www.tapiaconference.org
[2] https://cra.org/cra-w/discipline-specific-mentoring-workshops-dsw/

Sixty-nine students applied to the 2010 workshop and 40 were invited to attend. The students represented 21 institutions from 14 of the U.S. and Puerto Rico. In attendance were 24 African American students, 6 Hispanic students, and 10 students who identified as Other (Caucasian, Asian-American, Unlisted). Twenty-nine of the 40 participants were female. All but three students registered and attended the CHI 2010 conference after the workshop.

The first day of the workshop consisted of two panel presentations and two research topic presentations. The goal of the panels was to provide the students with insights on life in industry and academia as experienced by leaders in their field. Influential leaders including Jim Foley, Mary Czerwinski, and Mary Beth Rosson, provided their exciting perspectives on where the field had been, where it was going, and how to prepare their research for future success. These presentations were complemented with short "research highlight" presentations by several participating faculty members. The day concluded with an off-site dinner and networking opportunity at a local restaurant in Atlanta. The second day was filled with technical presentations from both academic and industry researchers on topics including Information Visualization (Jeffrey Heer), Privacy in Social Networks (Heather Lipford), Social Computing (Jason Ellis), and Interactive Information Search and Retrieval (Diane Kelly). These presentations were intended to expose the students to state-of-the-art research in various relevant HCI areas. The day concluded with a poster session and reception where the students presented their work in an informal reception setting and received feedback from peers as well as the participating academic and industry professionals.

The 2010 workshop was a tremendous success and was offered again in 2012 in conjunction with the CHI 2012 conference in Austin, Texas. The response was again fantastic. Out of 68 applicants, the workshop was able to support 25 students to attend. In addition, GRAND-NCE[3] funded five additional students from Canadian institutions. All 30 students were enrolled in Ph.D. programs representing 22 institutions in the United States and Canada. Participants included 10 African American, 6 Hispanic, 1 Native American, 8 Caucasian, 3 Asian American students, as well as 3 students who identified themselves as "Other." Twenty-two of the 30 total students were female. Five students reported having some form of physical disability (hearing, visual, or mobility) and one student with a severe mobility disability participated remotely from the University of Maryland. All but three students registered and attended the CHI 2012 conference after the workshop.

Based on feedback from 2010, several changes were made to the format of CHIMe 2012 to provide more opportunities for students and speakers to network in informal settings and to provide early opportunities for participants to learn about each others' research. In particular, two "Fast Forward" sessions for students and speakers were added and the closing poster session was removed. This workshop was also designed to run over one and a half days as opposed to two.

---

[3]http://grand-nce.ca/

Day one of the 2012 workshop (the half day) consisted of the fast-forward sessions and one panel entitled "Top Challenges/Directions in HCI." In this panel, experts including Ricardo Prada of Google, Juan Gilbert from Clemson University, Tessa Lau of IBM, and Beki Grinter from Georgia Tech, each gave their views on the fields most exciting research challenges and interesting directions for future work. This was followed by a research presentation by Tessa Lau. The day concluded with an off-site dinner and networking opportunity at a local restaurant in Austin. Day two consisted of several research presentations by industry and academic researchers including Enid Montague, Yolanda Rankin, Beki Grinter, and Cecilia Aragon. Day two also included two panels. The first was entitled "Former Student Panel–Success in a Graduate Program in HCI" and was moderated by CHIMe 2010 attendee Sheena Lewis. This panel consisted of several former CHIMe attendees and was a closed-door panel (no faculty or industry researchers present) desgined to encourage open discussion of the issues faced by students navigating a Ph.D. program. The second panel entitled "Industry, Government, and Academia" consisted of leading researchers from each sector, respectively, including Cecilia Aragon (U. Washington, LBNL), Enid Montague (U. Wisconsin), Mary Czerwinski (Microsoft Research), John Fernandez (Texas A&M), Ricardo Prada (Google), and Carlos Montesinos (Intel). The second day (and workshop) concluded with a panel entitled "How to Get the Most Out of CHI" and was designed to send the students off to a successful CHI conference with advice from repeat CHI attendees.

6.2.1   CHIME RESULTS

A survey of the CHIMe 2010 participants was conducted by the CRA one week after the workshop. The survey measured increase in participant interest, increase in confidence, and overall experience at the conference, among other factors. The results were overwhelmingly positive. Of particular importance are the impact that the workshop had on participant interest in the HCI field and confidence in their ability to complete a degree and conduct research in HCI. As seen in Figure 6.1, the workshop had a clear positive impact on these self-reported measures.

Perhaps even more telling than the findings from the survey data is the progress of the participants of the 2010 and 2012 CHIMe workshops. The organizers have informally tracked the original participants and located 57 of the 70 participants. Of these students, 4 are currently Ph.D. candidates, 4 are postdoctoral researchers, 21 are tenure track faculty, 23 are employed in industry or government positions, and 5 are in non-tenure track academic positions (lecturer, research scientist, director, etc.)

The workshop is also showing signs of sustainability. While there was a break after the 2012 offering, in 2018, the CHIMe founders, along with four previous CHIMe participants, organized the third workshop at CHI 2018 in Montreal.[4] The organization and execution has been successfully transitioned to a new generation of researchers and the funding for the workshop is now completely independent of the original CRA/CDC discpline-specific mentoring

---

[4]https://chime2018.wordpress.com/

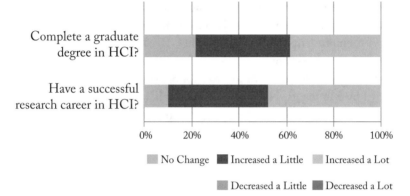

Figure 6.1: Results of the survey of the CHIMe 2010 participants.

workshop financial support. CHIMe 2018 was the most exciting offering to date with 32 supported participants and over 30 speakers from the CHI community.

6.2.2   LESSONS LEARNED

Organizing and running a mentoring workshop such as CHIMe is a serious undertaking. After being through the experience twice, the authors have learned several important lessons that will prove valuable to anyone attempting to replicate CHIMe in another research community.

First, co-location and cooperation with a major conference is critical. In all three CHIMe offerings, the organizers of CHIMe enjoyed a wonderful working relationship with the CHI organizing committee and support of the CHI conference chairs. While CHIMe was not executed as an official conference workshop in 2010 and 2012, CHI provided support in several critical areas. First, CHI provided space at no cost to the workshop organizers. Second, CHI provided administrative support to block register workshop participants for the CHI conference and to make hotel accommodations. The CHI organizing committee also provided an advertisement on the conference site to build awareness in the broader CHI community. In 2018, CHIMe enjoyed all of the usual benefits of being a true sanctioned CHI workshop. This co-location and cooperation is a win-win for CHIMe and CHI in many ways. First, it increases the diversity of the Ph.D. students in attendance while the participants get access to the leading venue in their research community. Second, co-location also means that many of the leading researchers will be available to speak to and network with the students during the workshop as well as throughout the conference. Finally, leading community researchers typically attend the CHI conference therefore additional funds are not needed to support travel for the speakers and panelists.

Second, financial support is critical because it provides the pathway for participants to not just attend the workshop, but just as important, to attend the conference that they should be attending as members of the research community. Student advisers, who must cover meals and lodging during the conference, are also incentivised as their funds are supplemented with workshop funds to support their student's travel. While the CRA DSW program provided nominal funds to support the first two workshops, additional fundraising was necessary. We have found industry and government partners quite willing to support these efforts. In fact, CHIMe 2018 was completely funded by Microsoft, NSF, SIGCHI, and SIGACCES.

Third, participation by the broader research community is crucial. The workshop success stems from community leaders' genuine participation in the sessions and interaction with the students. The small workshop setting, before the CHI conference, provides a safe and casual setting for students to interact with researchers who might be otherwise difficult to approach during the main CHI conference. Once the connection has been made, the students are then more likely to continue to interact with these leaders throughout the conference. This serves the obvious purpose of building the network of peers and mentors for the participants. It also serves to disseminate the student's research work to a broader audience.

Fourth, students benefit greatly from participants from both the majority and minority populations. Organizers should strive to include speakers and panelists from the majority as well as underrepresented populations.

Finally, cohort building is important in that it creates a small group with which each participant can experience the CHI conference. CHI is a large and potentially overwhelming conference. It is also a very social environment and one can easily feel left out of the community if attending alone. The workshop builds a small cohort before the conference even starts, giving the students a built-in social group to experience the conference with and to expand upon.

6.2.3   VISME: THE VISUALIZATION MENTORING WORKSHOP

How well does the CHIMe model translate to the VIS community? Is a successful VISMe workshop within reach? The authors are of the opinion that there is no real barrier to creating a successful VISMe workshop. While the VIS community is significantly smaller than that of CHI, this means only that the workshop would be smaller in scale as appropriate. The key to success of a VISMe workshop is not in organizing and execution—as the CHIMe model provides a blueprint. Rather, the key is in creating an inclusive environment in which participants feel welcome, get excited about the research community, and return from year to year. This can only be achieved by the genuine participation of the broader, inclusive VIS community in the effort. The strategies for building and maintaining that inclusive community are discussed in Chapter 7.

6.3   VISBUDDIES

Many conferences organize mentoring programs, aiming to pair newcomers to the conference with experienced attendees. For example, the SuperComputing Mentor-Protégé program that has run since 2008 matches an experienced mentor with a newcomer protégé (usually a student) from similar technical backgrounds; and comes with a set of obligations for both parties, including pre-conference email or phone contact and attending at least one event together. This type of matching can be very rewarding for both participants (as indicated by several testimonials), nevertheless there is a limit to how many protégés can actually participate, as potential mentors have multiple commitments during the conferences and thus the number of protégés volunteering can be larger than that of mentors.

In 2017, IEEE VIS community co-chairs Anastasia Bezerianos, Associate Professor at Université Paris-Sud, and Jonathan Woodring, Research Scientist at Los Alamos National Laboratory, organized VisBuddies, a more lightweight community building program. The goal of VisBuddies was to bring together new and returning IEEE VIS conference (VIS) attendees with similar interests, without the required effort of a full mentoring program.

Instead of pairing two attendees, as it was impossible to get a 1–1 ratio between experienced and new attendees, Visbuddies opted for a small group instead. They created groups of 5–6 people, mixing attendees with different levels of experience. The group size of 5–6 people ensured a small enough group to do activities together (eat, get coffee); and big enough that even if some members could not commit to common activities there would still be a critical mass for the remaining group to enjoy activities together. For newcomers, VisBuddies was a chance to meet experienced researchers in the field and discuss their research and conference experience. For returning attendees, it was a chance to meet fresh talent joining VIS for the first time, both students and young researchers as well as senior researchers and practitioners from other fields attending VIS for the first time.

In 2017, 350 participants (1/3 of the attendees of VIS) volunteered to participate in the initiative through the conference registration form. Their level of experience in attending VIS ranged from 0–20 or more years, with the majority (2/3 participants) being considered as newcomers/junior attendees (less than 2 years of attending VIS) and the remaining 1/3 participants considered as experienced/senior attendees (self-reported that they have attended the conference for 2 or more years).

6.3.1 MATCHING PROCESS AND RESULTS

Participants were matched automatically, through an algorithm that attempted to optimize the consistency of each group such that it included:

- at least two experienced participants (more than 2 years attending VIS);

- participants with common interests, as they were self-reported in the registration process (primary VIS and self-reported research keywords reported in the registration form), and

- participants holding a mix of positions, in both industry and academia.

The matching produced 60 groups of 5–6 people. The ratio of senior/junior attendees was roughly 2/3 per group (usually 2 senior and 3–4 junior attendees). Each group was contacted in a group email by the community co-chairs, and group members were encouraged to exchange information and meet in person at least once during the conference.

Participants were also encouraged to attend together the "Newcomers Meet-up," a pre-existing information session that gives newcomers tips about attending VIS. The community co-chairs co-organized the information session and prepared the room beforehand, placing labels for the different groups to help participants sit together with their group. These groups then went to lunch together. The community chairs were also in attendance to help with group assignment and make adjustments if needed. During the session, three people were alone in representing their group and were asked to join the closest group in terms of number (as group numbers indicate proximity in terms of research topic). People who attended the session without having participated in the VisBuddies program were invited to join the community chairs in forming additional ad hoc groups (that included the chairs). The presence of the chairs was an essential part of the process to encourage participants, help smooth the transitions to groups, and to ensure that everyone felt included in the process.

The community co-chairs were copied in email exchanges between 28 out of the 60 groups. All of these group exchanges led to plans for the groups to meet during the conference (and members of at least 18 groups met in the Newcomers meet-up information session—150 people in total). Several groups (at least 15) organized into multiple get-togethers with some of their group members. The majority of the remaining 32 (out of 60) groups very likely met as well, but the community co-chairs were not copied in their email exchanges.

6.3.2 OUTCOMES

Overall, this first VisBuddies initiative was clearly of interest to VIS participants given the number of people that volunteered to participate (approximately one third of the conference attendees). The community co-chairs have received informal feedback regarding the success of the initiative. Verbally and through email the chairs have been contacted by both experienced and newcomers to the conference to comment on the usefulness and the potential of the initiative for building a stronger community in the VIS conference. The willingness of participants to meet at least once and for several groups to meet again as a group, indicates the social value for the event.

The chairs also reached out to participants with an online questionnaire to collect feed-back on the event and gauge their experience and interest in participating in future VisBuddies offerings. The 57 responses received represents roughly 16% of the 350 people that participated in the event.[5] The majority of responses come from junior attendees (see Figure 6.2).

In VIS 2017 were you an experienced or a new attendee?
57 Responses

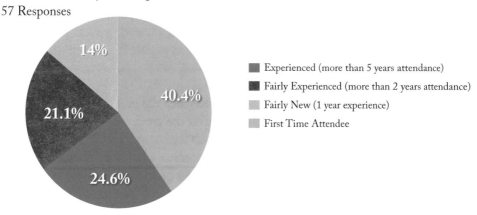

Figure 6.2: Experience of participants that responded to the questionnaire.

The goal of the questionnaire was three-fold: to understand what the participants felt they gained from the initiative, to determine if they maintained the connections they made due to the initiative, and to find ways to improve it.

Meetings: Participants were first asked if they did indeed meet their buddies, and if this only happened in the meetup. At least 55.6% (30 participants out of 57) of participants who responded actually met in the Newcomers meetup, indicating that having a dedicated venue aids with the logistics of meeting buddies. Another 44.4% (24 participants) mentioned that they met with their buddies more than once in the conference, indicating the matching can lead to connections that go beyond the structured event (Newcomers meetup) recommended by the initiative. The remaining participants either could not attend any meetings organized by their buddies (5.6%—3 participants) or did not organize to meet with their buddies (13%—7 participants). Although the questionnaire sample is small (16% of the initial event participants responded to the questionnaire) and likely biased toward participants who enjoyed the event, the responses indicate a confirmation that participants made an effort to meet with

---

[5]Fourty-four of the past 350 participants could not be reached as the email addresses used during the event were no-longer valid when the questionnaires were sent out. As several of participants were young professionals this can be explained by changes in their main institution.

their buddies (providing further evidences to our own observations of email exchanges).

Newcomers meetup: The vast majority of people responding to the questionnaire felt that the main value of the Newcomers meetup was as a means to meet their buddies (49.1%—28 out of 57 participants), although some still felt the event helped them find out more about VIS in general (21.1%—12 participants). Another 22.8% (13 participants) indicated that they could not attend the meetup because they had other engagements during that time—we note that several of these participants met with their buddies outside the meetup (given the overall numbers of people who met with their buddies). Finally, 10.5% (6 participants) did not find the event useful. Overall, these findings support that having a dedicated event to help people meet in person is important for most participants, but it is good to pair it with secondary objectives (e.g., advice for newcomers to the conference).

Further contact: Although less than a year has passed since the event ran, we were curious to see if participants kept in touch with their buddies after the conference. We were delighted to find that 21.4% (12 participants) actually kept in touch with some people from their group (a further 1.8%, 1 participant, said they acknowledge each other when they meet). The remaining 76.8% (43 participants) have not had any contact with their buddies since. Results can be seen in Figure 6.3. We also asked participants if they plan to meet any of their buddies in the upcoming VIS 2018 conference. The majority (42.6%—23 participants) stated that if the opportunity arises (e.g., they run into each other) they would be open to it. Another 1.8% (1 participant) had already made plans to meet with their buddies. The rest (29.6%—16 participants) said they have not considered it (and a further 24.1%—13 participants are not attending). Results can be seen in Figure 6.4. Combined, these results indicate that although in some cases the initiative

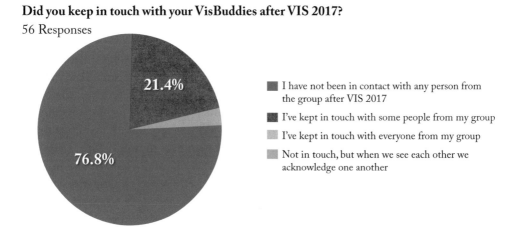

Did you keep in touch with your VisBuddies after VIS 2017?
56 Responses

- I have not been in contact with any person from the group after VIS 2017
- I've kept in touch with some people from my group
- I've kept in touch with everyone from my group
- Not in touch, but when we see each other we acknowledge one another

Figure 6.3: Contact after VIS.

helped make connections that went beyond the VIS 2017, forming lasting relationships takes time and requires several occasions where people can interact. One potential solution is to offer lightweight initiatives to help people keep in touch, such as sponsored alumni lunches or ways to communicate when former buddies are attending conferences.

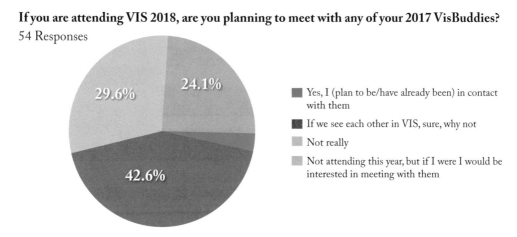

If you are attending VIS 2018, are you planning to meet with any of your 2017 VisBuddies?
54 Responses

- Yes, I (plan to be/have already been) in contact with them
- If we see each other in VIS, sure, why not
- Not really
- Not attending this year, but if I were I would be interested in meeting with them

Figure 6.4: Plans to meet with previous VisBuddies at upcoming VIS.

Matching success and benefits: On a Likert scale, participants indicated whether they felt the matching of the initiative was successful as well as to which benefits they felt the initiative brought them (remaining items in Figure 6.5). As the responses indicate most participants felt the matching was fairly successful in bringing together people with similar research interests and different experiences. When it comes to the benefits there does not seem to be one single benefit that stands out apart from the fact that it made the attendance experience more enjoyable for many participants. The other reasons (e.g., meet peers, meet more experienced researchers, form connections) seem to resonate for a subset of the participants.

Joining again: When asked if they will participate in the initiative in following years, 25.4% (14 participants) responded they would definitely attend again and 40% (22 participants) that they are considering it. On the other hand, 7.3% (4 participants) responded that they did not find it useful and thus would not attend again. A further 25.4% (14 participants) stated they would not attend the conference.

Other feedback: Free form comments from questionnaire participants indicated that one aspect that could be improved would be to somehow ensure the commitment of more senior people (who could not always join the groups), and to adjust what is considered as "senior" to not include students (in our matching definition we considered as senior anyone with more than two years

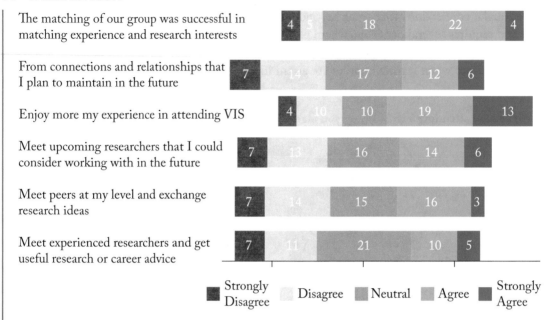

Figure 6.5: Success of the matching algorithm and effects of the VisBuddies experience.

experience). Getting senior people to commit to meeting with juniors and participating in the event is a true challenge given their time constraints and general overloaded schedules.

6.3.3 FUTURE DIRECTIONS

Although the VisBuddies program seems to have been appreciated by participants there are clearly several aspects that could be improved. First, some participants were surprised they were labeled as senior/experienced in their group, and others were disappointed by the lack of truly senior people in their group. The setup of the automatic matching algorithm frequently resulted in senior people that were older Ph.D. students with multiple years in the field, mainly due to the number of participating attendees with extensive professional experience (e.g., professors) being small. While considering alternative thresholds for distinguishing experience/seniority is important, there is still potential value in senior students serving as mentors given their availability compared to the non-student senior members of the community.

Nevertheless, ensuring the involvement of more senior researchers and practitioners in the field is clearly needed. This is an important challenge given that conferences are opportunities for seniors to meet with colleagues, resulting in very busy schedules. It is crucial that the community as a whole appreciates the need for (and benefits that come from) mentoring newcomers in the field.

Although several email exchanges between group participants showed that participants were planning to attend similar research sessions (indicating their interests were aligned), other aspects of the matching algorithm can still be improved. In particular, data collected from the participants is being analyzed to determine other aspects/properties that could be included to improve the matching.

It seems the Newcomers meet-up session was not attended by almost half of the participating attendees based on the email exchanges. The reason for this is currently being investigated. Even though the questionnaire did not reveal a specific reason, primary feedback indicates this may have been an issue with the timing of the session (it took place before the main conference started). Nevertheless, it is also possible that several groups prefer the freedom of meeting outside a prearranged event.

6.3.4 DIVERSITY AND INCLUSION

How can VisBuddies be used to foster diversity and inclusion at VIS? As noted earier in the chapter, mentoring is a critical component of inclusion and thus retention. While not designed as a purely mentoring intiative, VisBuddies clearly includes key elements of mentoring, including the ability of newcomers to network and meet experienced attendees as well as the opportunity to learn from experienced attendees how to get the most out of the vis conference. Ensuring that participants from underrepresented populations are distributed across matched groups is one way to begin to promote diversity. A potentially more effective option could be to match senior members of VIS with junior members of underrepresented populations, to help juniors gain targeted mentoring. These two are somewhat competing goals from an algorithmic perspective given that the number of senior and junior attendees from underrepresented groups is low. To truly help promote both inclusivity and diversity one could consider branching into both, a more targeted mentoring program for juniors in the field, and a more social initiative to help attendees meet other people in the same field. VisBuddies currently sits in the middle of these two goals. In both cases, one of the biggest challenges is to engage and convince senior participants to contribute to the process, as they remain a key aspect of any mentoring and networking effort. Existing events (such as panels on diversity and inclusivity) can act as safe places to communicate to both seniors and juniors about the needs of our community and to promote mentoring events.

6.4 CONCLUSION

In this chapter, we have presented two examples of efforts to engage diverse participants in the CHI and VIS research communities, respectively. While these are just two examples, they provide a blueprint for successful engagement, and evidence for the effects of such efforts in broadening participation in a research community. Such engagement is critical to retention and thus critical to diversity and inclusion efforts.

CHAPTER 7

Building Inclusive Communities

Johanna Schmidt, Kelly Gaither, Mashhuda Glencross, Michelle Borkin, and Petra Isenberg

7.1 INTRODUCTION

As has been alluded to previously, inclusion reflects an individual's feelings and sense of belonging. It equates to respecting and appreciating everyone's diversity [90]. While it encompasses a complex set of factors, we can think of inclusion in our context as social inclusion and academic inclusion. Christa Freiler states that, "social inclusion gets to the heart of what it means to be human: belonging, acceptance and recognition [81]." She also provides an important, but often omitted perspective of what it means to exclude. In the extreme, exclusion most often impacts vulnerable individuals or populations, those less valued. Academic inclusion is defined in education as approaches to educating all learners together. If we look at this in the context of higher education, it has a great deal of direct relevance to us as a community. By examining the contrary, academic exclusion refers to the denial of academic opportunities for some groups or students [82]. This denial can happen in a number of ways, including bias, harassment, and long-term exposure to seemingly small infractions or microaggressions[1] (frequently referred to as death by a thousand cuts [91]). *Inclusive communities* encourage and ensure the participation of all groups, and make sure that all individuals can explore their differences in a safe, positive, and nurturing environment. They enable all individuals to function at full capacity, where acts of exclusion and injustice based on group identity and other factors are not allowed to occur, or to continue. An inclusive community [83]:

- does everything that it can to respect all its citizens, giving them full access to resources, and promoting equal treatment and opportunity;

- actively works to eliminate all forms of discrimination;

- engages all its citizens in decision-making processes that affect their lives;

---

[1]Microaggressions are defined as slights and putdowns based on race, gender, gender identity, age, sexual orientation, disability, culture, national origin, language, ethnicity, religion, and socioeconomic status that can make a group marginalized.

- values diversity; and

- responds quickly to racist and other types of discriminating incidents.

    Building inclusive communities is not an easy undertaking and requires persistent commitment to ensure progress and growth, but there are a number of inclusion and diversity initiatives in place today across industry and academia, from which we can draw inspiration [84]. We can start with the philosophy that inclusion and diversity are necessary ingredients to a strategy for excellence. This requires that we, as a community, embrace the notion that inclusiveness and excellence are interdependent, rather than separate concepts. Building an inclusive community is achieved by a series of investments in programs, policies, procedures, and plans that in aggregate create an environment that is diverse, welcoming, and inclusive for all members. As a starting point, we provide strategies and best practices by examining existing efforts toward this end, many of which the VIS community has begun (Section 7.2). Based on principles provided in "10 Simple Rules" for achieving gender balance, we then analyze efforts that are underway and propose directions for future efforts (Section 7.3).

7.2   STRATEGIES FOR INCLUSION

It is worthwhile to look at proven strategies for inclusion when defining guidelines for building inclusive communities. If we think about inclusion within the context of organizations and communities, and we recognize that VIS is, in and of itself, a community, we can discuss a number of approaches and best practices that draw from industry, government, and academia. Given that medicine sits at the interface of science, technology, and people, it is no surprise that a number of health-related institutions have spent significant time thinking about these issues. MD Anderson lists ten characteristics of an inclusive organization as [85].

- **Accepting diversity and inclusion as a way of life**—Diversity and inclusion exist at all levels and all members come together to coordinate action toward achieving common goals.

- **Evaluating individual and group performance on the basis of observable and measurable behaviors and competencies**—There is a clear understanding of roles, responsibilities, and expectations and they are achievable.

- **Operating under transparent policies and procedures**—There are no hidden rules of behavior or sets of expectations known to some and unknown to others.

- **Maintaining consistent interactions with everyone**—There is no double standard; rules are applied evenly and appropriately throughout the organization.

- **Creating and maintaining a learning culture**—Career development is encouraged and supported. Mentoring programs are robust. Mistakes are recognized and viewed as learning opportunities rather than character flaws.

- **Providing a comprehensive and easily accessible system of conflict resolution at all levels**—Conflict is inevitable. Transparent systems are in place to address this conflict in a fair manner that respects dignity and confidentiality of all parties.

- **Recognizing that it is part of the community that it serves**—Members of the community should serve as active participants in community activities, playing a vital role in addressing needs.

- **Living its mission and core values**—An organization that promises one thing and delivers another risks losing the trust and confidence of its constituents.

- **Valuing earned privilege over unearned privilege**—Members are recognized for their actions and accomplishments, not simply their titles or degrees. All are treated with respect regardless of their status or class.

- **Accepting and embracing change**—Change is inevitable. Current and past practices must be reviewed and updated continuously to meet the changing demands of its constituents.

While these characteristics may seem self-evident and common-sense in nature, they can be difficult to implement in a manner that ensures buy-in and longevity. However, there are a number of interesting initiatives affront that we present to provide the reader with tangible examples of successful programs that have been implemented.

7.2.1 ORGANIZATIONAL DIVERSITY AND INCLUSION

A number of professional organizations have started to focus on building diverse and inclusive communities. This is being realized at the macro or top level and and the micro or in-depth levels. There are a number of factors to consider when trying to integrate inclusion and diversity into the fabric of the organization.

Organizational Oversight Responsibilities

To maintain an ongoing commitment to diversity and inclusion, it is necessary to create mechanisms and lines of responsibility at multiple points in the hierarchy. Corporations and academic institutions appoint officers whose job it is to grow, monitor and maintain diversity and inclusion over time. Virtual organizations, including professional organizations, draw from these efforts and address issues that are relevant to them. The Computing Research Association's Committee on the Status of Women in Computing (CRA-W) developed and published a set of best practices for running an inclusive conference [92]. They list steering committee opportunities and responsibilities as follows

- **Balance demographics of steering committee members**—This balance should be done along a number of dimensions including seniority, topical expertise, geography, academia/industry, gender/racial/ethnic balance, disability status, and academic. Additionally, efforts to avoid cronyism should be put in place to ensure a good mix of long-term expertise with the infusion of new people.

- **Maintain institutional memory**—Collect and maintain longitudinal data on metrics relevant to conference diversity and inclusion and track trends over time.

- **Make diversity and inclusion a key priority**—Appoint conference organizers with diversity and inclusivity as key goals. Avoid assigning all "housekeeping" roles to women and underrepresented groups. Create a position on the organizing committee responsible for diversity and inclusion.

The SuperComputing (SC) conference series [86] recognizes that increasing diversity and inclusion won't happen unless they take an active role in affecting change. To this end, they appointed Diversity&Inclusion Chairs to oversee issues including child care, reasons for attendee dropoff, cultural barriers, code of conduct, reviewer bias, committee climate, attendee climate, and many others that contribute to growth and a feeling of inclusion. ACM SIGGRAPH, ACM SIGCHI, IEEE VIS, SIGGRAPH, CHI, and VIS, recently appointed a diversity chair. The role of the chair is to lead and collaborate with a team of dedicated volunteers to develop a diversity and inclusion strategy aligned with the organization's mission. The strategy will take actionable steps that work toward integrating inclusion-related topics 2018 and content into the SIGGRAPH Conferences and that provide information and best practices for building a diverse and inclusive organization for its members. In SIGGRAPH in Vancouver held its inaugural Diversity Summit. This day-long event took a first step in providing informative presentations and panels on the subjects of global diversity and inclusion, and in tackling issues such as implicit bias. A follow-on summit is set for SIGGRAPH 2019.

Organizational Ethics and Code of Conduct

Whether or not an organization is legally required to have a code of conduct, it is beneficial for every organization to have one [87]. It serves to clarify an organization's mission, values, and principles and articulates standards of professional conduct and desired behavior. It also serves to signal that "good citizenship" in the field is an important component of the behavioral ethics of an organization [89]. Further, research suggests that organizations with rigorous codes of conduct have a higher public perception of good citizenship, sustainability, ethical behavior, and social responsibility performance [88]. They are instrumental in reinforcing constructive behaviors, which is increasingly relevant in organizations with multicultural constituents. Further, a strong code of conduct goes beyond awareness of expectations and offer mechanisms for reporting.

A number of professional organizations have codes of conduct, including IEEE and ACM. Additionally, a growing number of professional conferences have begun developing and publishing a code of conduct that articulates their values and expectations. IEEE VIS has recently gone through the process of developing a strong code of conduct that includes both expectations and mechanisms for reporting. In line with the IEEE code of ethics [103], the IEEE VIS code of conduct is as follows.

> IEEE VIS is committed to providing an inclusive and harassment-free environment in all interactions regardless of gender, sexual orientation, disability, physical appearance, race, or religion. This commitment extends to all IEEE VIS sponsored events and services (webinars, committee meetings, networking functions, online forums, chat rooms, and social media) and any interaction regardless of affiliation or position. As a community that aims to share ideas and freedom of thought and expression, it is essential that the interaction between attendees take place in an environment that recognizes the inherent worth of every person by being respectful of all. IEEE VIS does not tolerate harassment in any form. Harassment is any form of behavior intended to exclude, intimidate, or cause discomfort. Harassment includes, but is not limited to, the use of abusive or degrading language, intimidation, stalking, harassing photography or recording, inappropriate physical contact, and unwelcome sexual attention.

> Anyone who experiences, observes, or has knowledge of threatening behavior is encouraged to immediately report the incident to ombuds@ieeevis.org. All information shared will be kept confidential. In cases where a public response is deemed necessary, the identities of victims and reporters will remain confidential unless those individuals consent otherwise.

> IEEE VIS reserves the right to take appropriate action to foster an inclusive and respectful environment. Attendees violating these rules may be asked to leave the conference without a refund, at the sole discretion of the conference organizers. In addition, attendees are subject to the IEEE Code of Ethics.

Contact:

- Please contact ombuds@ieeevis.org if you experience, observe, or have knowledge of behavior in violation of the Code of Conduct.
- Please contact inclusivity@ieeevis.org with any questions about the Code of Conduct and Inclusivity & Diversity at IEEE VIS.

Organizational Efforts to Mitigate Bias

Peer review is the cornerstone of the scholarly publication process. In the simplest terms, experts in a given field are asked to weigh in on the merits of a given work or set of works, their identities are hidden, and it is designed to encourage peer impartiality. In an ideal circumstance, peer review provides "a system of institutionalized vigilance" [93]. This is well regarded as a fair process when reviewers provide impartial evaluations, and we rely on this process being fair and unbiased. However, more often than not, we hold these processes without recognition that bias takes many forms and is introduced by both implicit and explicit means. Mitigating this means that we acknowledge that bias occurs and put mechanisms in place to mitigate detrimental effects [92].

- **Ensure broad demographics along many dimensions**—Including seniority, topical expertise, geography, academia/industry, and gender/racial/ethnic balance.

- **Encourage double-blind reviews**—Research shows that it is effective at mitigating conscious and unconscious bias in the review process [94]. McKinley provides additional valuable insight into possible bias in the review process and the merits and implementation of a blind/double-blind review [95].

- **Provide unconscious bias, harassment, and ethics online training materials and ensure that reviewers go through the training**—As reviewers, we are subject to demonstrate a number of biases that we may not be aware of—content-based, confirmation, conservatism, bias against interdisciplinary research, and publication bias [96]. We are also guilty of demonstrating prestige bias, language bias, affiliation bias, and gender bias. These may be introduced maliciously, but more often than not, they enter through our inability to recognize our unconscious biases.

Building a diverse and inclusive conference committee has very strong benefits with respect to implicit bias, which ultimately affects content selection for conferences. Experience drawn from building the General Submissions jury for SIGGRAPH 2016 and 2017 demonstrated that balancing diversity requires significant commitment from the beginning of the process while seeking committee members, partly because many of our professional networks are not very diverse, and our natural inclination is to reach out to those we know and are familiar with. To mitigate this, SIGGRAPH employed the following process.

- Seek recommendations from key volunteer community members.

- Seek wider volunteers through an open submission system.

- Seek volunteers through social media.

- Balance diversity of expertise.

- Balance diversity of selected members.

- Send invitations for jury in batches.

- Re-balance until expertise and diversity requirements met.

This process had two objectives: (1) to widen the pool of volunteers and committee members and (2) to balance demographic and experiential diversity of its members in a 100-person review committee. This required an iterative process. Following this structured process made it possible to build diversity into a 100-person review team. SC follows a similar process [86].

Organizational Programs to Increase Diversity and Inclusion

As demographics have shifted over the years from a fairly homogeneous population to one that is increasingly diverse, more and more organizations are putting programs in place that recognize this diversity. Inclusive organizations understand that it is important to not only put programs in place that draw people in, but also to mitigate known barriers to entering and remaining in the organizations. These programs and efforts to increase diversity and inclusion include the following [92].

- **Onsite childcare or available resources for local babysitters and mechanisms for reimbursement**—Corporations have long understood the value of providing childcare to attract and retain talented employees. Academia has been slower to recognize this, but perhaps the slowest to evolve has been conferences. These meetings provide a vital forum for academic researchers, facilitating networking, communication, collaborations, and contact with funding agencies. Early-stage researchers are most impacted by a lack of onsite childcare, leaving parent-researchers to choose between attending key conferences and furthering their career and caring for their children [97]. This impacts an organizations ability to increase diversity and becomes a self-fulfilling prophecy, making it impossible to "move the needle."

  In place since 2015, the VISKids program provides childcare grants during the IEEE VIS conference (VIS). The grants are available for VIS attendees with small children that need to bring them to the conference, or who are forced to incur extra expenses by leaving their children at home. The grants nominally cover expenses for childcare, and VIS attendees make their own arrangements. In addition to the grants, VISKids also organizes meetings throughout the IEEE VIS conference for people to gather and discuss, for general information, and a welcome meeting at the beginning of the conference [98]. The concept started at VIS 2014 as an effort to organize a childcare room where kids could go to play (under the supervision of an accompanying adult), breastfeeding mothers could pump milk, and diapers could be changed. That initial effort gave rise to a "family room," but legal issues prevented officially advertising it formally through the conference. Through persistent efforts to chip away at the impediments, VIS volunteers made this a reality, balancing a fine line between legal constraints imposed by the sponsoring organization and the obvious needs of young researchers to balance professional growth with the demands of being a caregiver. As an alternative, SC provides onsite childcare [86] through the use of KiddieCorp, an organization that provides onsite childcare services for conferences and convention centers [99].

- **Travel support for attendees**—Increasing participation from underrepresented communities requires initiatives to increase attendees by drawing from outside the normal attendee pool. This oftentimes mandates that attendee expenses be reimbursed or at the very least, greatly reduced. A number of initiatives in this space have centered on scholarships of one form or another that seek to substantially reduce or if possible, remove the financial impact of attending.

VIS initiated an Inclusivity & Diversity Scholarship in 2018 to increase participation by people of all backgrounds and identities and from underrepresented or historically marginalized groups. The scholarship is need-based and gives preference to first-time attendees seeking to learn more about research and development activities at VIS. Additionally, recipients are assigned mentors to welcome them to the community and to serve as an extended network beyond the conference itself [100].

• **Consider accessibility issues**—People with disabilities are often omitted from the conversation of diversity and inclusion, but one billion people or 15% of the world's population have some form of disability. One-fifth of the global population have significant disabilities. Barriers include inaccessible physical environments and transportation, unavailability of assistive devices and technologies, communication, gaps in service delivery, and discriminatory prejudice and stigma in society [101]. A number of countries legally mandate accessibility for persons with disabilities, but others lag further behind. Corporations recognizing that inclusion and diversity mean recognizing and including those with disabilities have led the charge in initiatives to not only include, but formally recognize that we must all recognize accessibility as a vital piece of the inclusion conversation [102].

7.3   ANALYSIS OF STRATEGIES AND OUTLOOK

We can measure progress in our community by the litmus in the *Ten Simple Rules* defined by Martin [104]. As a community, we are waking up to the fact that this is an important effort, one in which we can and should play a role. More importantly, we have a responsibility to capitalize on the fact that we speak a "universal language," well versed in the visual arts and rooted in interdisciplinary work. We spend a great deal of our time at the boundary between computer science and another science or engineering discipline. This makes us uniquely positioned to take a stand and as a community, respond as leaders rather than followers. To this end, we measure our progress and point out those items that require further action and examination.

• **Rule 1: Collect the Data**—It is difficult to know what the diversity picture looks like without collecting data. While this can be somewhat complicated, corporations globally have started to take the initiative. Moving the needle can't happen without knowing a baseline, and that baseline is directly informed by counting. For professional organizations, this means creating mechanisms for counting when members register, allowing them to opt in.

• **Rule 2: Develop a Speaker Policy**—This is directly tied to Code of Conduct and defines acceptable behavior for speakers and speaker conduct. Additionally, this gets to the fact that diversity and inclusion should be considered when choosing invited speakers, encouraging the choice of diverse speakers when relevant and appropriate.

- **Rule 3: Make the Policy Visible**—There is no substitute for transparency. All policies should be made available so that the entire community can glean what is and more importantly, is not, acceptable. This serves as a signal to both established members of the community and newcomers the ethics and boundaries of a given organization.

- **Rule 4: Establish a Balanced and Informed Committee**—Achieving true diversity on the steering, executive, and program committees is a long-term commitment and requires patience and determination. However, truly achieving diversity on these committees will pay off in the long term, encouraging young researchers to participate and become future leaders.

- **Rule 5: Report the Data**—Reporting data is only secondary, and may be considered parallel, to collecting data. Publishing progress, good or bad, signifies that the effort is truly important to the organization.

- **Rule 6: Build and Use Databases**—Maintain information in a database such that information is preserved longitudinally. This allows information to be preserved as the organization experiences multiple changing of the guards.

- **Rule 7: Respond to Resistance**—It is difficult to speak up in an environment in which you perceive that you are unwelcome, but it is important nonetheless. The ability for members of an organization to speak truth to power gets at their ability to be heard. An organization is only as healthy as its most silent members, and it is incumbent upon an organization to encourage all voices be heard.

- **Rule 8: Support All Voices at Meetings**—Members of underrepresented communities oftentimes find themselves in positions of service, where their voices are represented by other members or simply unheard. Support for those voices that are being overshadowed or drowned out is crucial to the health of the organization long term. Even small infractions can be addressed by other members when they see them by explicitly calling on those whose voices are being quieted.

- **Rule 9: Be Family-Friendly**—The recognition that one's professional life is only a fraction of one's whole life means that we recognize that our members have other issues that pull at their attention. If we consider and encourage our members to be well-rounded and we respect their entire identity, we open our membership to not only our traditional participants, but also those from more diverse communities.

- **Rule 10: Take the Pledge**—Making a strong and vocal commitment to diversity and inclusion cannot be overstated. While it may take time to achieve the desired results, a public and persistent commitment does wonders to signal to underrepresented communities that this organization is truly dedicated to welcoming them, to making a change, to embracing diversity and inclusion long term.

CHAPTER 8

Marshalling the Many Facets of Diversity

Bernice E. Rogowitz, Alexandra Diehl, Petra Isenberg, Rita Borgo, and Alfie AbdulRahman

8.1 INTRODUCTION

Diversity is not a goal unto itself. Increasing the count of people from different genders, races, and geographies is important, but the real goal is to create communities that are substantially better because their diverse members share benefits and opportunities equally. To achieve real inclusion requires sharing responsibility, prestige, recognition, and power. We have learned from natural ecosystems that diversity is critical for sustainability and health, and that different populations need to contribute to the collective gene pool [171]. A healthy academic ecosystem depends on the introduction of new ideas and connections, and on the diverse voices that carry them.

Many factors contribute to the diversity of an academic ecosystem, such as the diversity of topics and disciplines, the gender, racial, and geographic make-up of its membership, and the dynamics of international research funding. There are many ways to measure individual empowerment, such as authorship, leadership, and recognition. In this chapter, we examine a range of diversity vectors through the lens of the IEEE VIS family of conferences, and explore how these interact with measures of recognition. To do so, we have analyzed data on many facets of our organization and its participants. We explore how the evolution of topics has created opportunities for the inclusion of new academic disciplines, and how topic diversity can promote gender diversity. We analyze how our members become empowered through their participation as authors and program committee members, and through awards for technical achievement. We examine patterns of international research and development spending as a backdrop for understanding the factors that contribute to geographical diversity, in general, and in our ecosystem, specifically. Our goal is to provide insight into the diversity of our ecosystem and how it has evolved, and to increase awareness of the sociological factors that underpin our future success.

Chapter Overview. This chapter focuses on sociological factors influencing diversity in visualization. Our observations are based on the analysis of data about the IEEE VIS family of conferences (IEEE Visualization (IEEE VIS), Information Visualization (InfoVis), Scientific Visualization (SciVis), and Visual Analytics Science and Technology (VAST)), and on data from external sources.

The field of visualization grew primarily out of computer science and data-rich experimental sciences, and the population characteristics and traditions of these disciplines still play a dominant role. Since the 1990s, the field has grown to include a wider range of disciplines and research topics, enriching the scientific ecosystem. The first section of this chapter looks at the growth and structural evolution of the IEEE VIS family of conferences and symposia, and at the evolution of topics revealed through the analysis of keywords used in papers and in calls-for-participation in the various conferences. We look at census data to shed light on the relationship between topic diversity and gender diversity. Our hypothesis is that new topics can expand our diversity by attracting scientists from different disciplines, which may have very different intrinsic gender distributions. To get a handle on this, we present data on how the proportion of male and female professionals has evolved in different disciplines over time and identify opportunities for increasing diversity by embracing different disciplines. We observe, for example, that the proportion of women professionals in computer science is steady, or decreasing, yet, the proportion of VIS program committee members from computer science is increasing.

Next, we examine gender diversity in the leadership of our community. To set the stage, we looked at the gender diversity of our authors over time, examining number and proportion of authors who are women, and also, the proportion of papers with at least one female author. Both these measures have increased over the past 30 years. To get a measure of the degree to which male and female members are valued and esteemed, we then looked at two measures of recognition. First, we studied the make-up of the program committees of the various conferences. Only members of the community with excellent credentials and judgment are invited to serve on the program committees, since their main job is to evaluate the conference manuscripts, which later become archival publications of the *IEEE Transactions on Visualization and Computer Graphics* (TVCG). We also looked at the gender-distribution of awards, which are our society's way of recognizing technical achievement, and discuss the relationship between participation and recognition in our community.

The IEEE VIS family of conferences has broad international participation, integrating diverse intellectual and cultural experiences into our community. In this section, we examine sociological and financial factors that drive funding for international research and diversity programs. We look at R&D funding across a wide range of countries, discuss different funding patterns in developed and emerging countries, and explore programs through which the richest countries support international research and diversity. We explore different funding patterns in developed and emerging countries and its impact on research publication in peer-reviewed journals. We also look at how research outlay influences the flow of researchers around the world, show the growth of international participation in VIS program committees.

This chapter, thus, looks at diversity from a number of different angles and perspectives, to examine how topics, participants, and geography interact to affect our ecosystem. We conclude with observations, suggestions for improvement, and aspirations for the future.

Data Caveats and Limitations.   Visualization-specific data in this chapter revolves around the IEEE VIS family of conferences and symposia. There are other important venues such as the EG/VGTC Conference on Visualization (EuroVis), the IEEE Pacific Visualization Symposium (PacificVis), and the IS&T Visualization and Data Analysis Conference (VDA), plus many journals serving the visualization community. Data about their topics and participants would help expand our understanding of diversity in visualization.

Much of the data in this chapter was painstakingly culled, by hand, and small errors may have been introduced. The IEEE has not tracked authors, program committee members, or recognition by gender, race, ethnicity, country of origin, or seniority. We scraped data from past programs and calls-for-papers. We collected gender information based on first names, personal knowledge, searches of web pages, use of his/her pronouns in posted biographies, and photos on LinkedIn. We specify where we were not able to determine gender from these sources. Although we recognize the importance of respecting gender identity, we did not have access to clarifying metadata. We did not address race and ethnicity directly, since we were not able to compile data on how individuals self identify. Although the population of underrepresented minorities (URMs) in visualization is very low, it would be important to understand their representation among authors, committee members and award winners. Data on geography, funding, and migration were extracted from governmental and private websites. We looked at country of origin for program committee members, largely by combing through online biographies. There are enormous subtleties in their collection and curation that may have eluded our scrutiny. Many of our analyses are not specific to visualization. Through our analysis of funding programs, we can provide some indication of how funding is being allocated to gender diversity, worldwide, but we do not have data on how well these programs have done in driving a more diverse research population. Our hope is that our work will help frame this discussion of diversity within a larger sociological context, and will provide motivation for creating more complete data sets, which will enable more sophisticated analyses.

8.2   TOPIC DIVERSITY AT IEEE VIS

IEEE Visualization was launched in 1990 [162] and has changed significantly over its history, not only in size, but also in diversity. Inspired by this 25th Anniversary, Isenberg and her colleagues created a database of its published papers and authors, plus a search tool, and two papers analyzing topic evolution [155, 156]. This section draws on their work, and also explores topic diversity through keywords used in the calls for the major conferences.

Figure 8.1 shows the evolution of the IEEE VIS family of conferences, from a small conference with 55 papers in 1990 to a symposium with 3 major tracks and 7 specialized symposia

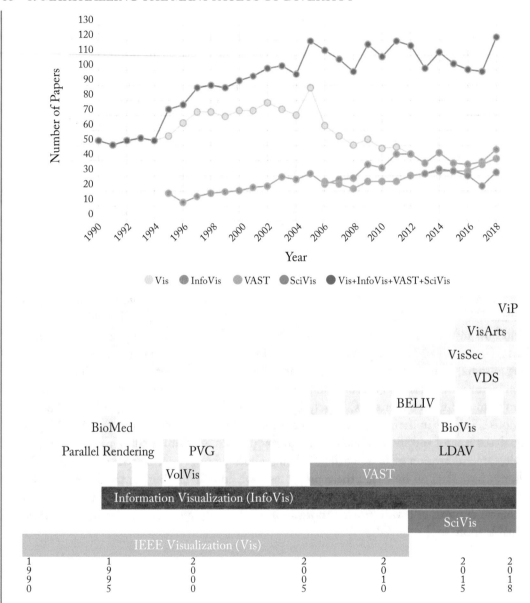

Figure 8.1: The evolution of IEEE Visualization family of conferences. The top panel, from [155] shows the steady growth in technical papers in the major VIS conferences since the inception of IEEE Visualization (VIS) in 1990. (This image is in the public domain.) The bottom time chart, updated from that paper, depicts changes in the major conferences and also depicts the expanding set of symposia associated with the conference.

in 2018. Over that time, the number of papers has almost tripled. In 2015, the symposium published over 120 papers, which all appeared in the *Proceedings of the IEEE Transactions in Visualization and Computer Graphics* (TVCG). In 1995, there was a major expansion, with the introduction of InfoVis and two scientific visualization symposia, Volume Visualization (VolVis), and Parallel Rendering, which evolved into Parallel and Large-Data Visualization and Graphics (PVG). In 2006, VAST was launched, supporting the visual analytics community. Several major symposia and workshops have developed since. BELIV was introduced in 2006 to address the evaluation of visualizations. In 2011, the BioVis conference emerged from a series of conference workshops on biological data visualization dating back to the 1990s, and the LDAV symposium on Big Data Analysis and Visualization arose to address the astounding growth in large-scale data. In recent years, there has been an upsurge in new symposia, including Visualization for Cyber Security (VizSec), the Visualization Arts Program (VisAp), and Visualization in Practice (ViP).

In their 2014 paper, Isenberg et al. [158] remarked on the intrinsic diversity of the visualization field, its roots in many disciplines, in the research methods it embraces, and in the application areas it explores. It is clear just looking at this structure, that the major conferences, and perhaps especially, the associated symposia, draw from a large pool of scientists and practitioners from diverse disciplines.

The diversity of topics has evolved. In the early 1990s, the main focus was on algorithms and scientific visualization. The creation of the Information Visualization conference in 1995 reflected a change in focus, welcoming new topics and participants from adjacent disciplines, such as perception, human computer interaction, and statistics. VAST provided introduced new topics in visual analytics and modeling, and welcomed disciplines where visualization is used as an analysis methodology. The biology, art, and cybersecurity symposia expanded the envelope of disciplines, enabling the growth of new topics such as gene expression analysis, aesthetics, and network analysis.

Two analyses give us a deeper look into the evolution of topics at VIS. Isenberg et al. examined keywords for the ~4300 papers submitted to the annual IEEE VIS meeting, including InfoVis, SciVis, and VAST. They found a significant increase in "interaction techniques," and "evaluation," plus in keywords related to "time-varying data" and "multidimensional/multivariate data," including "machine learning" and "statistics." The author-generated keywords that declined most were those relating to "volume visualization," "meshes, grids, and lattices," as well as "numerical methods/mathematics." It is possible that papers in these areas have been subsumed by the associated LDAV symposium.

Using data from the KeyVis database [155], Figure 8.2 illustrates the time-course of four topics over the span of our history. We see, for example, "interaction" and "evaluation" first appeared in paper titles in the early 2000s, and have been gaining momentum since, reflecting the growing emphasis on human-computer interaction. Popular topics in scientific visualization, such as "isosurface" and "volume visualization," have declined.

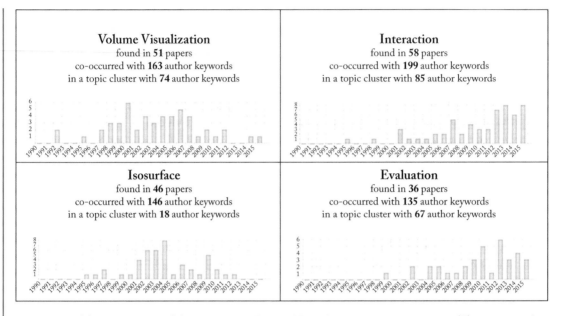

Figure 8.2: The evolution of four topics at IEEE Visualization, 1990–present. These examples reflect the ebb and flow of diversification and extinction over time.

To complement this analysis, we examined raw text from conference calls-for-participation for the three major IEEE VIS conferences in the most recent decade, from 2009–2018. Word-frequency clouds for individual words in these texts are shown in Figure 8.3. Info-Vis, VAST, and SciVis are shown in the rows. We stratified the data into two time periods as a way of identifying major changes over this period. For InfoVis and VAST, the two intervals are 2009–2013 and 2014–2018. Since SciVis was launched in 2012, the first word cloud contains two years of keywords, from 2012 and 2013. These word clouds share a common scale, so that the counts across conferences are preserved. For example, "studies" was mentioned 20 times in the InfoVis calls from 2009–2013, "methods" was mentioned 20 times in VAST in that same time frame, and thus are the same size. The SciVis 2012–2013 data have been scaled up proportionately. Common words such as "visualization," "data," and "information" were removed from all visualizations to better reveal the fine structure.

The conferences have distinct flavors, with InfoVis focused on user studies, evaluation, and design, VAST focused on analytics and representation, and SciVis on "science," "hardware," and "devices," "perception," and "interaction." Some topic shifts, cross-referenced against the actual counts in the data, can be observed. Some InfoVis topics that appeared frequently in 2009–2013, such as "mathematics" (8 mentions) and "interaction" have dropped out in the most recent 5 years; other topics, such as "context" (15 mentions), "analysis," "design," and "integration" have appeared, or have grown significantly. This rotation shows topic diversity and evolution. VAST

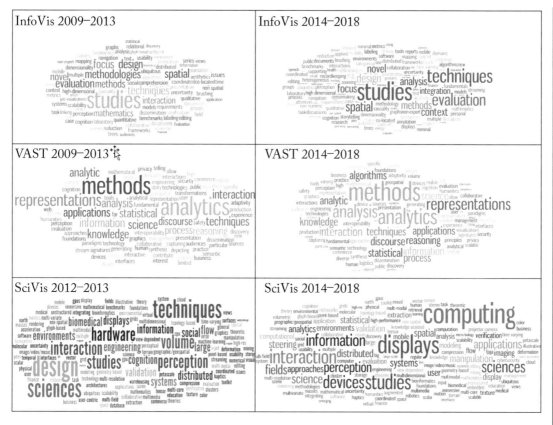

Figure 8.3: **Call-for-Papers Text Analysis.** Word clouds show high-frequency keywords for VIS conferences over the most recent decade, divided into two time intervals. InfoVis and VAST are separated into two 5-year intervals, from 2009–2013 and from 2014–2018. Since SciVis was created in 2012, its first interval covers 2012–2013 only.

terms have not changed much over the past decade, but this may, however, reflect the re-use of text year over year, not stagnation in the topics being addressed. Using the same methodology, of iteratively filtering out terms that appear equally in both periods, the strongest difference was the increase in the term "algorithms" which grew from 5 mentions to 10. In SciVis, cornerstones of the 2012–2013 period, such as "hardware," "techniques," and "volume-rendering" have all but dropped out, new terms such as "computing" (26 mentions) and "displays" have become prominent, and new topics such as "cybersecurity" and "robotics" have appeared.

The above analysis is a first step toward understanding the dynamics of topic diversity and evolution in visualization. It would be fascinating to look at other visualization conferences, explore richer datasets, and conduct more sophisticated analyses. For example, we would like to study the co-located conferences and workshops at VIS, since they seem to bring enormous diversity into our ecosystem. On the analysis side, Isenberg et al. [155] clustered visualization papers into 186 categories based on several keyword types associated with each paper. It would be fascinating to examine how topic clusters have formed, morphed, and declined over our history, and relate these dynamics to changes in the demographics of our population.

8.3   TOPIC DIVERSITY CAN DRIVE GENDER DIVERSITY

In this section, we look at topic diversity as a possible on-ramp to population diversity. We focus on gender diversity because we were able to find relevant data that bear on this question. The basic idea is that some intellectual disciplines may have intrinsically higher proportions of women, underrepresented minorities, and international researchers, so, by embracing these disciplines, we include more diverse populations in our ecosystem. Our analysis focuses on professional women, since this is the only group for which we were able to obtain sufficient data.

Figure 8.4 shows the proportion of women professionals in disciplines closely allied with visualization, such as computer science, compared with their participation in fields that are related to the emerging topics we have described. These data are extracted from Nathan Yau's Flowing Data site [170], which joins two sources of labor data from the Census Bureau (1950–2000) and from the American Community Survey (2010 and 2015), coded according to the 2010 ACS job classification system. The top-left panel shows how the proportion of women has varied in professions most closely related to Visualization fields. Roughly 25% of computer programmers (red) and computer and systems analysts (yellow) are women, and there has been a sizable drop in female programmers since 1990. There there has been a consistently high rate of female statisticians (purple), with representation steady around 50% since 1980. The graph in the bottom left panel shows womens' participation in fields that have more recently been integrated into the visualization community. Social sciences (blue), medical and life sciences (green), and art (magenta) have nearly equal participation by men and women, and these numbers have been consistent over the span of measurement, which in many cases reaches back to 1950.

We often hear language suggesting that the low participation of women in scientific visualization is a direct consequence of the low participation of women in the "hard sciences." The fields shown in the top right panel tell a more nuanced story. While, indeed, female participation in physics and astronomy (violet) have made very slow increases from their low levels half a century ago, there has been a steady increase in female participation in chemistry and materials science (azure). The job category that includes environmental science and geography has experienced a remarkable leap in female participation, rising from less than 10% in 1950 to 40% at the most recent measurement in 2015.

The chart in the lower right quadrant of Figure 8.4 distills these data, showing average participation rates for women since 1990, when IEEE VIS was launched. Each color-coded bar corresponds to a discipline in one of the other graphs. To explicitly compare computer science fields with the disciplines, their colors (red and yellow) are more saturated. Also, a red horizontal

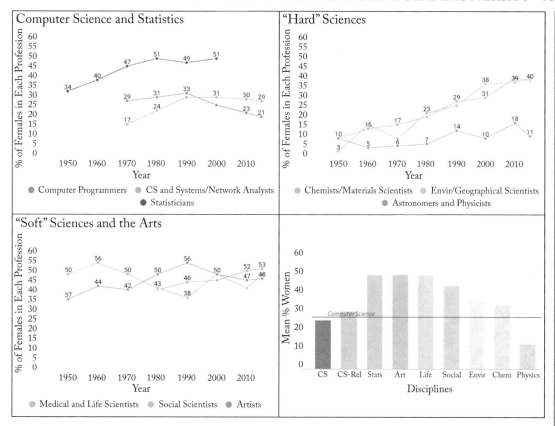

Figure 8.4: The proportion of women in fields relevant to Visualization. Three panels show the growth of women professionals in "computer science and statistics," "soft sciences and the arts," and "hard sciences." Data since 1990 are summarized in the bottom right panel. Women make up 25% in computer science disciplines (red and yellow bars) compared with roughly 50% in statistics, the arts, life and social sciences. In chemistry/materials and geography, 35% are women. Only physics/astronomy has a smaller proportion of women professionals than computer science.

line at just over 25%, depicts the average proportion of women in these two fields. The representation of women in computer science fields over the past 30 years is half that of their proportion in statistics, art, and life sciences, which are all around 50%. The proportion of women in the social sciences is also near parity. Women professionals make up nearly 35% of environmental scientists, geographers, chemists, and materials scientists. Across all these fields, the proportion is lower in just one category, physics, and astrophysics.

The decreasing proportion of women in computer science is a danger sign for the growth of gender diversity at VIS. If the number of female computer science graduate students is not increasing, then continuing to draw members from that pool will not contribute to increasing the female/male ratio of our population. We did a short analysis to understand how this decrease may affect the visualization ecosystem. We compiled background on the disciplines of VIS program committee members. Figure 8.5 shows the proportion of computer scientists, sampled every five years from 1995–2015, plus the two most recent years, 2017 and 2018. Two decades ago, roughly half the program committee members were computer scientists. That proportion has grown steadily since, and is currently near 80%. So, not only is the pool of female computer scientists decreasing, but our program committees are increasing drawing from that pool. If our goal is to attract a more diverse population, it's clear that we need to encourage participation from fields outside of computer science.

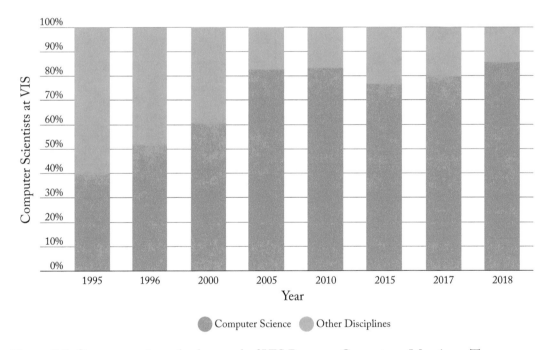

Figure 8.5: Computer science background of VIS Program Committee Members. The percentage of computer scientists has increased from about 40% to over 80% since 1995, during which time, the proportion of women in computer science has been falling.

We see topic diversity as an important lens through which to examine growth drivers for diversity in visualization. Many of the disciplines examined in Figure 8.4, not only dovetail with new and diverse topic areas for visualization, but could also increase the participation of women, since these professions have higher proportions of women scientists and practitioners. Our community has already embraced many of these topics, and they have been an important vehicle for keeping ideas fresh and responsive. For example, the enormous growth in data for biological and genetic analysis has attracted doctors and biologists to visualization. Statisticians, geographers, and social scientists are increasingly using visualization to analyze the vast pool of geo-located social data now coming available. The important goal of providing evaluation methods and guidelines for visual representation have drawn new members to our community from psychology and human computer interaction fields. The arts program has attracted artists as well as other professionals interested in visual representation, semiology, and expression. New topics add to our diversity, and also attract practitioners from other fields with higher intrinsic proportions of women, and perhaps other minority groups, creating a virtuous cycle.

Looking forward, why not develop visualization symposia that explicitly tap disciplines with higher proportions of women scientists? For example, the EnviroVis symposium at EuroVis draws on disciplines related to the environment and geography, and there are definitely large data analysis and representation issues in chemistry and materials science that could benefit from visualization. Topics could also be promoted that tap visualization opportunities for data journalism and advocacy, attracting social scientists, graphic artists, and writers.

8.4   GENDER AND RECOGNITION AT IEEE VISUALIZATION

A hallmark of a healthy social ecosystem is one where individual merit is acknowledged and recognized. In this section, we take a look at three levers of intellectual recognition through the lens of diversity. To begin this exploration, we examine the number of women authors at IEEE Visualization. Next, we look at gender diversity in program committee composition and in recognition through awards.

Women Authors at IEEE VIS.   The number of papers at VIS has been increasing steadily. An analysis of all VIS paper authors from 1990–2016 is shown in Figures 8.6 and 8.7. These data were coded by hand, and include as "unlabeled" the 1 or 2 authors per year for whom we were not able to definitively ascribe gender.

The number of women authors rose quickly from 8 in 1990 to 37 by 1996. In the following decade, however, the number of women authors leveled out (solid red line), while the number of male authors grew monotonically (dotted red line). In the most recent decade, the rate of male authorship has continued to increase 50% per decade. During this decade, the rate of women authors has doubled, owing largely to participation in InfoVis and VAST. Although the number of women authors is still small, the recent growth rate points to growth in ecosystem diversity.

Program Committee Membership.   Program committee members' main responsibility is the review of technical papers. This is a very competitive process, with roughly 20% of the papers accepted, and these accepted papers appear as full publications in the *Transactions on Visualization and Computer Graphics* (TVCG), a highly prestigious journal of the IEEE Computer Society. Program committee members provide in-depth technical reviews, solicit additional reviewers, and adjudicate over often-conflicting reviews. Program committee members are recognized for

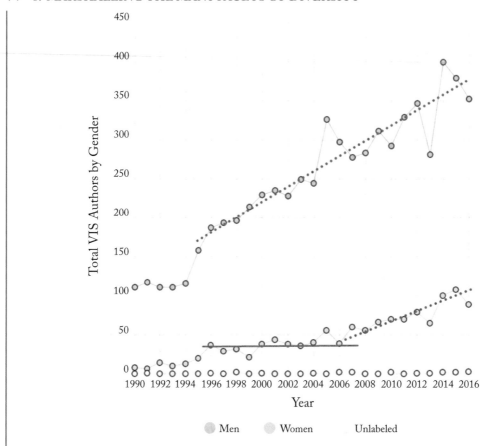

Figure 8.6: Men and women authors at VIS. The number of authors has risen steadily over the history of VIS. The number of male authors has increased at a rate of roughly 20 per year since 1996. The number of female authors began growing steadily in 2006.

their excellent credentials, judgment, and knowledge, and play an integral role in maintaining the quality and intellectual integrity of the organization.

Figure 8.8 shows the male vs. female composition of the program committees for the major conferences, sampled irregularly from 1995–2018. We are missing data from the first five years, but from 1995–2005, women made up roughly 10% of the program committees. This number jumped to just under 20% in 2006, driven by the steadily increasing proportion of women on the InfoVis program committee. The overall proportion is now over 20%, driven by the continued increase in female participation in InfoVis, and a significant growth in SciVis, as well. Figure 8.9 explores these data more closely, plotting the percentage of women on the program committee as a function of the percentage of women authors, for the same years from 1995–2016 depicted

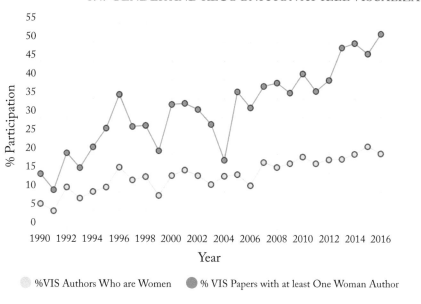

Figure 8.7: Proportion of women authors and papers with women authors. The proportion of papers with at least one woman author has increased faster than the proportion of authors who are women, reaching 50% in 2016.

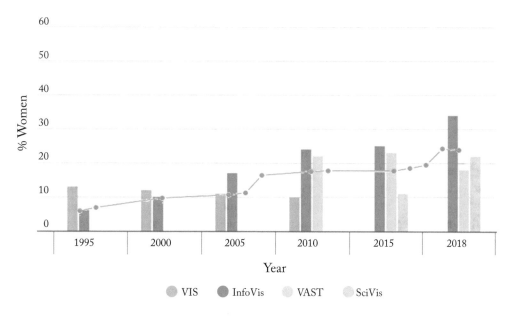

Figure 8.8: Program committee composition. On average, the percentage of women serving on program committees has grown monotonically (red line), led by the InfoVis conference.

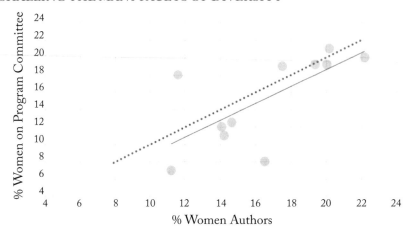

Figure 8.9: PC representation of women relative to authorship. Participation on a program committee is related to authorship (r-squared = .475), shown as the black line. The red line, however, shows the prediction if being selected for a PC were at parity with authorship.

above. The correlation between the percentage of women authors and the percentage of women on the program committee is 0.475 (r-square). That is, authorship is related to leadership at VIS. However, the proportion of women on the program committee is roughly 2% lower than would be predicted by the proportion of women authors (dotted red line). That is, the rate that women are invited to a leadership position on a program committee does not keep up with their level of scientific contribution.

Society Awards.   Each year, the IEEE and individual conferences provide recognition of intellectual achievement through an annual awards process. Awards serve as a mechanism for validating members' value in an organization, and add to their prestige and influence. Figure 8.10 shows the recognition structure at VIS, comparing how frequently men vs. women receive awards for their intellectual contribution to the community. In the 14 years from 2004–2017 there have been yearly awards for Technical Achievement and for Career recognition. During this 14-year period, the annual Technical Achievement Award was awarded to a man every year, but one. Up through 2017, no woman received the Career Award. Although not included in this graph, 2018 marked the first year that a woman was bestowed a Career Award, raising the proportion from 0–6.7%.

Starting in 2013, the three major conferences, InfoVis, SciVis, and VAST have awarded best paper awards. These awards reflect our current award dynamics, not a reminder of behavior in decades past. Of these 15 awards, 3 papers with female authors have been recognized. For the first time ever, in 2018, two papers with at least one woman author were awarded a best

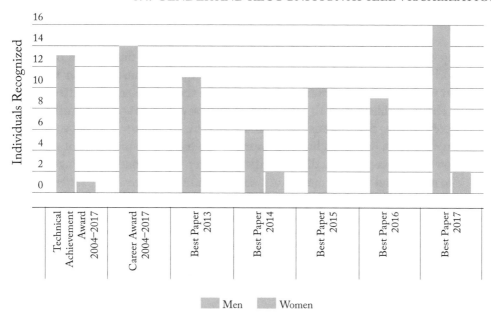

Figure 8.10: Awards Recognition. Recognition by men and women via conference-wide Technical Achievement Career awards and Conference-specific Best Paper Awards. Recognition of women is far below their representation as authors.

paper award, bringing the percentage from 8–20%. However, of the 54 authors who have been recognized by a Best Paper award, only 4 have been women (7.4%).

Across all these opportunities for intellectual recognition, there have only been five instances, counting 2018, where a woman received a major honor. There are certainly reasons why this finding may not reflect bias, such as the longevity of women in the field, the number of students they may have had to contribute to their success, etc. Still, the absence of yellow on this chart is breathtaking.

Since 2014, the proportion of women authors has grown to 20%, and as has the proportion serving on program committees (PCs) for the major conferences. Although this is not a large percentage, this growth reflects growing recognition. The same cannot be said for the awards process. A much more thorough study would be required to delve into the many factors that drive recognition in a society. This is important to explore, because participation and recognition make people feel respected, acknowledged, and admired in an organization, giving them authority, voice, and status. Also, awards and recognition provide valuable line items on resumes, which can increase the chance of getting a job or a grant, which can also have economic implications. We hope that these data will spark discussion and awareness.

8.5   GEOGRAPHICAL DIVERSITY AND FUNDING

Another important facet of diversity in visualization is geographic diversity. In the Arts, Philosophy, and Politics, differences between cultures are explicitly considered. In Visualization, too, different world experiences contribute differently to the field. Rene Descartes and Jacques Bertin were French, William Playfair and James Clerk Maxwell, who sculpted the first 3-D visualization, were Scottish, S.S. Stevens and John Tukey were American, Herman von Helmholtz was German, to name a few. Each brought a very different flavor to the fabric of visualization.

So, how is that geographical diversity driven? A step toward understanding geographical diversity is to understand the distribution of research funding. Governments differ significantly in terms of the amount of money they allocate for research, which has a strong influence on the magnitude of research activity. Figure 8.11 shows the distribution of gross domestic expenditure on research and development R&D. The countries with the biggest R&D budgets are the U.S., China, the European Union, Japan, and Germany. The amount spent by these countries dwarfs the investment by other developed and developing countries. The chart to the right re-plots these data as a pro portion of overall GDP, providing insight into the Research appetite for each country. In order, the countries with largest expenditure in R&D relative to their GDP are: Israel, South Korea, Japan, Switzerland, and Sweden. The color-coding in the graph shows that the countries with the largest R&D budgets (red) are not necessarily the ones with the highest expenditure relative to GDP, and vice versa.

Richer, more developed countries can support more research and can attract students from all over the world. As we will see, some are more generous toward foreign students, and others spend their research funding to support their own populations. Developing or emerging countries can offer fewer opportunities, and students often leave to study in richer environments. Whether they stay in their adopted countries or return home, this exchange expands scientific borders and increases diversity.

8.5.1   THE FLOW OF INTERNATIONAL RESEARCH FUNDING

To better understand how research funding is allocated across the world, we collected public data on agencies and universities that provide grants to support international research and diversity programs. Since governmental funding agencies in each country drive the research agenda, the fields they choose to fund may differently affect funding for visualization research. In richer countries, the diversity of funded topics is very large. We included grants that are targeted for specific countries as well as opportunities that are open to all international researchers. This compilation is not comprehensive; it is intended to provide a glimpse into the magnitude and scope of programs that support international collaboration and diversity.

Figure 8.12 provides a high-level overview of our findings. This tree-map shows international-focused research funding and diversity programs for the countries that devote the most money to research and development. Rectangle size represents total R&D budget, color-

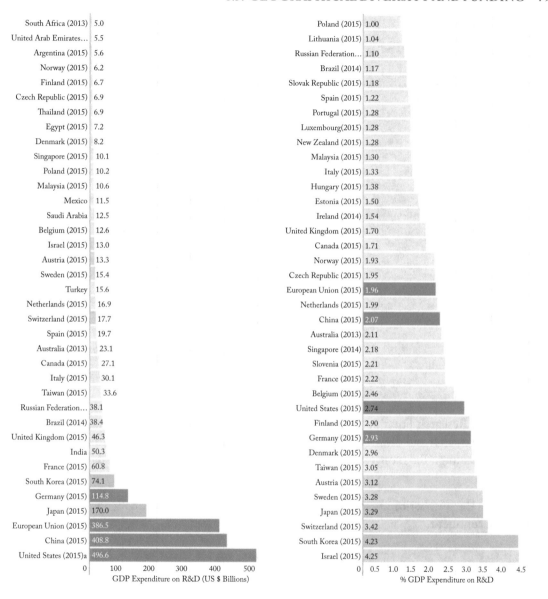

Figure 8.11: Research Funding by Country. The chart on the left shows R&D funding by country, with the US topping the charts at $500B, and China close behind at over $400B. The right-hand chart depicts R&D funding as a proportion of gross domestic expenditure (GDP). Countries with the highest absolute spending on R&D are shown in red; countries with the highest R&D expenditure relative to GDP are highlighted in blue. Countries with highest absolute spending *and* highest spending on R&D relative to GDP are shown in purple. Data come from the 2018 NSF report [132].

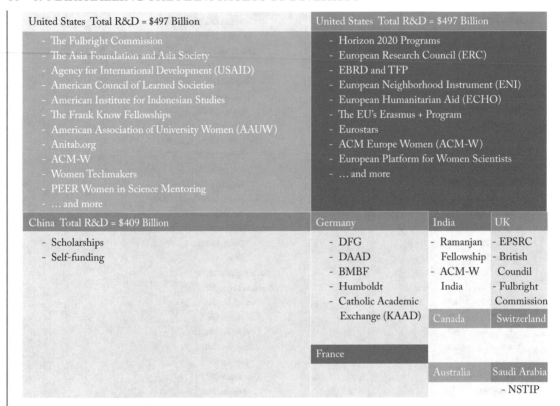

Figure 8.12: Research Funding for International and Diversity Programs. A sample of the international research and diversity programs funded by the richest countries. Area is proportion to R&D spending. The more research programs we were able to discover, the darker the color.

ing depicts the number of international or diversity programs we identified, and title bar color codes the continent.

The U.S. and The European Union.   In this figure, we see that the U.S. and the European Union are the largest contributors to worldwide research, and this wealth translates into a wide array of programs to support funding for developing countries, international collaborations, and diversity. The U.S. has the largest R&D budget. Major funding goes to support international collaboration and research in developing countries, including joint collaboration initiatives between Asia and the U.S. (such as the ASEAN Research Program), and between Africa and the U.S. There are also many programs supporting diversity, including the National Academy of Sciences PEER Women in Science Mentoring Program [140], plus non-profit international organizations such as Anita Borg [141], Women Techmakers [142], and the ACM Council on Women in Computing (ACM-W) [143].

The European Union offers many funding opportunities including Horizon 2020 [163], the European Research Council, and the European Neighborhood Instrument (ENI). There are several programs supporting research between Asia and Europe, such as the EUforAsia Programme, the Trans-Eurasia Information Network (TEIN), and the Asia-Europe Foundation (ASEF) as well as the partnership programs between Africa and Europe [167]. Recently, the European Commission (EC) opened its funding to any country of the world that wants to apply [144]. In 2021 they will launch Horizon Europe, with a budget of 100 billion euro, the biggest research budget in history. On the surface, this initiative seems to be a great opportunity for non-European countries, but is not clear how much it will cost to participate, and what proportion of funding will be allocated outside the EU.

Other Developed Countries.   Individual countries in Europe, especially Germany, Switzerland, and the UK, also have active programs supporting gender equality, research collaborations and student fellowships in developing countries [131]. Germany has the largest R&D budget of the European countries, and is one of the strongest contributors to research funding in developing countries. The Deutsche Forschungsgemeinschaft (DFG) has specific cooperation programs for 90 countries in Africa, North and South America, Europe, and Oceana. Moreover, the DFG strongly supports gender-equality programs [147]. In the UK, the Engineering and Physical Sciences Research Council (EPSRC) funds programs for International collaboration with China, India, Japan, and the U.S. [148]. They also support programs that fund equality and diversity, including gender, place of origin, and other factors. Since the announcement of Brexit, the UK has suffered significant losses in funding from EU projects, such as Horizon2020 [166]. An inflection point will come in 2019, when the UK leaves the EU, jeopardizing its right to participate in the research budget.

Emerging Countries.   During this past decade, emerging economies such as Saudi Arabia and China have significantly increased their engagement in the international academic community. This shift can be understood within the context of a new theory on the Economics of Innovation [168], in which countries are changing their economic paradigm from trade- or oil-based to knowledge-based [151]. In this context, countries are investing in increasing their research and patent portfolios and bootstrapping their research programs. Their main mechanism is to send their students to the U.S., Europe, and Australia for advanced degrees. China has the second largest research and development budget, worldwide, which they focus mainly on the development of their citizens. China invests in scientific programs for its students, including international research visits for undergraduates and support for Chinese students to earn their masters' degrees in the U.S. and the EU. From 2001–2011, China has increased funding for research from 1.0–1.8% of GDP, with a target of 2.07% in 2015 [149]. A recent report from the National Science Foundation (NSF) [132] shows a direct correlation between the investment in science and the quantity of published papers. Figure 8.13 shows the increase in peer-reviewed papers in the European Union, the U.S., and in China. The results are breathtaking. China's

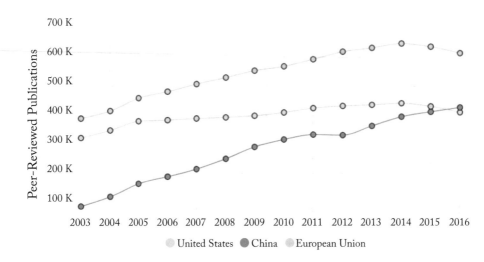

Figure 8.13: Peer-reviewed publications for the European Union (EU), China, and the U.S. Data are courtesy of NSF 2018 Report [132].

output has quadrupled since 2003, and in 2018, it surpassed the U.S. as the country with the most peer-reviewed papers.

In Saudi Arabia, the National Science, Technology and Innovation Plan (NSTIP) is dedicated to increasing scientific publishing and patents. To do so, Saudi Arabia recruits researchers to institutions such as the King Abdullah University of Science and Technology (KAUST). Its main investment is in sending Saudi students abroad for post-graduate study, to increase their knowledge and their international networks. India, Japan, and South Korea have also been increasing their R&D expenditure [149].

Other developing regions such as Latin America vary greatly depending on the economic fluctuations on the region. We found one regional effort, FRIDA [159], that funds Digital Innovation in Latin America and the Caribbean. Also, there are private efforts made by companies such as Google, Microsoft Research, and Facebook to increase the mobility of undergraduate and Ph.D. students from Latin America to the U.S. and Europe.

8.5.2   PATTERNS OF MIGRATION

Abel and Sander [160] analyzed migrants demographics to understand contemporary trends in international migration. They identified some interesting trends that are relevant to our analysis of international diversity in research, including: (1) the attractiveness of North America; (2) significant movement from South Asia to the Gulf states; (3) diverse flow dynamics within Europe; and (4) North America and Europe as the principal flow sinks; Asia, Africa, and Latin America as the main sources of migration.

Many of these patterns can be explained by the funding dynamics we outlined in the previous section. Wealthy regions like North America and Europe invest in high-quality graduate education that attracts students from Asia, Africa, and Latin America. This, coupled with fellowship and scholarship programs, provides a powerful magnet. Iconic programs such as the German Marie Sklodowska-Curie Fellowships [152], the American Fulbright scholarships [153], and the English IAESTE [154], for example, promote mobility of researchers in the early stages of their careers, independent of their age and country. Likewise, countries that see education as a strategy for growth, such as Saudi Arabia and China, have generous programs that provide funding for their citizens to study abroad.

Mobility is very important for international collaboration, research exchange, and diversification of ideas. The literature investigating mobility patterns among scientists and Ph.D. students is vast [169]. Recently, Bohannon [165] analyzed a set of 741,000 public CVs to understand migratory patterns of scientists. This research was based on the ORCID datasets [164], an important collection of worldwide scientists' profiles, collected by Dryad, a nonprofit repository for scientific data. Although the data are incomplete, comprising only 10% of scientists profiles in which Europe is overreprsented, the results are intriguing. For example, 15% of the scientists in the dataset had migrated.

International Participation at VIS.   It would be interesting to study migration patterns in VIS, where it seems, anecdotally, that there is a growing population of international students who are graduate students or post-docs in the U.S. and Europe, especially in Germany. Figure 8.14 shows our first attempt in this direction. These data show the country-of-origin of program committee members for the three IEEE VIS conferences, irregularly sampled from 1995–2018. Data were hand-coded based on personal knowledge and on information gleaned from individuals' CVs.

Looking at the left figure, we see that VIS has always been an international conference, with at minimum 16 countries represented in the program committees. This number grew in 2010, and has been increasing steadily since. Although this is a 60% increase, in many cases, the country is represented by one person. The graph to the right shows the growth rate by country. Here we show the countries with the highest growth rates plus the U.S. for comparison. There has been a 20% increase in PC members from China since 2000, an 8% increase in participants from Austria, and moderate growth in the other countries shown. By contrast, although scientists born in the U.S. comprise 60% of the program committees, their growth rate is near zero.

8.6   SPRINGBOARD FOR FUTURE DIRECTIONS

Studying an ecosystem is a difficult task, since there are many dimensions, influences, and interaction effects. Taking this first step has shown us how much we don't know. One big factor in tackling the unknown is the lack of data. IEEE, for example, could capture more information

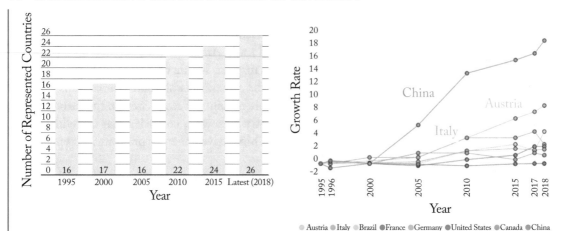

Figure 8.14: Geographic distribution of PC members. The left-hand graph shows the number of different countries-of-origin of VIS program committee members. The right-hand graph illustrates the growth rate in PC participation for selected countries.

about VIS participants, such as the conferences and symposia they attend, their demographics (e.g., race, gender, ethnicity, where they were born, trained, and work), and their discipline of study. This repository could be joined with information the IEEE already has about participants' membership, publication/presentation history, leadership roles, and awards. The society could also mine funding information provided by authors, which could give us a handle on the geographic flow of funding specifically focused on visualization research.

More ambitiously, we think it would be valuable to understand the generation of new ideas in our ecosystem, and what drives their introduction. Do these new ideas germinate because we have a vibrant system that rewards and encourages diversity? Looking at the calls for papers over the last decade, we did find topic evolution. However, we had to look hard to find dramatically new ideas and directions. Our hunch is that a lot of new and innovative ideas at VIS are coming from the panels, workshops, symposia, and meet-ups, which seem to be bursting with new topics and ideas. If this is true, is it because there's a lower barrier to entry for papers in these venues, widening the entry portal? Or, is it because they explicitly draw from fields outside of computer science? Are they more gender- and race-diverse? And if so, can this diversity be tied to differences in the demographic populations from which they draw? These events are part of the fabric of VIS, and contribute to its appeal and success. Explicitly studying their role in encouraging diversity in visualization would be an interesting way to explore some of the hypotheses raised in this paper.

We also need to build reward mechanisms that will encourage diverse voices. Fellowships and travel grants could encourage women and minority scientists and build our geographic footprint. Awards committees could be asked to explicitly add people from diverse backgrounds to the pool of candidates. Conference and symposium chairs could specifically ask their leaders to be on the look-out for scientists and practitioners who can bring diverse perspectives to our program and organizing committees. Even small honors can have a big impact, such as asking someone to chair a session or serve on a committee. That first invitation can lead to many new opportunities for individuals, and bring cascading insights into the organization.

8.7   CONCLUSION

This chapter looked at the IEEE Visualization community through an ecological lens. Diversity is critical for maintaining the health of an ecosystem, and to do so, all members need the opportunity to thrive. We examined diversity statistics for women authors, program committee members, and award recipients, and studied several exogenous factors that affect the diversity of our community, such as the disciplines and home countries of our population. We also explored the role of multidisciplinary applications and symposia in broadening our population demographics, widening the opportunity for inclusion. We have seen a healthy increase in the topics and disciplines in our current mix, and see exciting opportunities for growth. We are happy to show a steady increase in the geographic distribution of our program committee members. But we have work to do. Although the proportion female authors and program committee members has grown over the years, the numbers still hover around 20%, and the number of women who have won recognition for their academic achievements is embarrassingly small. The representation of African American and Hispanic scientists is almost uncountably low. In the VIS 2017 Panel on Diversity in Visualization, two young women scientists shared their lived experience in our community, showing how much further we need to come in simply treating members who have different backgrounds with respect. Visualization as a field has thrived on cross-pollination from diverse disciplines and perspectives, increasing the demographic and geographic diversity of our community. We hope the explicit links we have drawn between diversity and ecosystem health will help guide our vision for the future.

8.8   ACKNOWLEDGMENTS

This chapter was inspired by the 2017 Diversity in Visualization panel, and especially by Aviva Frank's insights on the sociology of diversity. The first two authors would like to thank Rita Borgo and Alfie Abdul-Rahaman, who collected and pre-processed data about the IEEE VIS conferences. Petra Isenberg generously shared her curated collection of VIS data and provided critical contributions. We also thank our students Gemza Ademaj and Natkamon Tovanich who helped with the daunting task of collecting and cleansing data for our analyses.

CHAPTER 9

Future of Diversity in Vis

Ron Metoyer and Kelly Gaither

"Society as a whole benefits immeasurably from a climate in which all persons, regardless of race or gender, may have the opportunity to earn respect, responsibility, advancement and remuneration based on ability."—Sandra Day O'Connor

Although the definition of underrepresented groups may change over time and as a function of global geography, there will undoubtedly always be underrepresented populations in a given community. While we have provided an overview of the sociological constructs, barriers, and issues facing various types of underrepresented communities, the concepts are not necessarily tied to the demographics or a local definition of underrepresentation. Rather, the concepts and constructs are applicable generically. Additionally, building and supporting communities that engage all voices, that foster innovation, and that celebrate differences is an effort shared by all and is an effort that will undoubtedly pay significant dividends over time.

As we saw in Chapter 1, it is in the best interest of the community and the country to diversify the workforce and to begin to create inclusive communities in which all members of the workforce can thrive. While diversity and inclusion are related concepts, they are not the same. It is not enough to simply create a diverse community without addressing inclusion as well. In Chapter 2, we were introduced to the larger social context in which these concepts of diversity and inclusion live and why they are necessary. In particular, it is critical to understand that power structures perpetuate the inequities that fight against diversity and inclusion and that they must be balanced to produce a more inclusive environment. Chapter 3 presented specific barriers that push against diversity and inclusion in many forms. It became clear, in this chapter, that there is no simple, clean solution, but rather that diversity and inclusion must be at the forefront of our minds (all of our minds) throughout every facet of the research community. This is especially true for those who operate from a position of privilege within the community.

Chapter 4 took an uncomfortable look into our colleague's experiences, allowing us all to see what can and does happen when someone encounters a non-inclusive environment. This chapter allows us all to feel the impact that those experiences can have on one's life and career. The hope is that through this chapter, we all can develop empathy, to see the effects of our actions.

Chapters 5–7 took a more prescriptive look at building a diverse and inclusive community. In particular, Chapters 5 and 6 presented specific examples of programs such as BPVis and CHIMe, designed to recruit and retain a diverse community of researchers. The hope is that these examples provide models for and spark new ideas for initiatives within the IEEE VIS community. In Chapter 7, we encountered specific strategies and policies that can begin to promote diversity and address the power structures within the community that counter inclusion. We must be diligent and unapologetic as we implement best practices such as the CRA-W recommendations for an inclusive conference or the Ten Simple Rules for measuring progress. The strategies and practices in this chapter simply must be put in place and enforced in order to make quick progress in creating an inclusive community.

Finally, in Chapter 8, we took a broader look at diversity, opening the lens much wider to understand the impacts of topic diversity and geographic diversity in the context of diversifying VIS. Through informal data analysis, we were introduced to interesting patterns and trends across gender and geography and presented with concrete opportunities for the IEEE VIS community to address some apparent biases. These are just a sample of the ways in which we can examine diversity outside of underrepresentation to understand how inclusive environments can encourage creativity and shared advancement.

Now is the time for change. As a community, IEEE VIS has an opportunity to lead this change and serve as a beacon to other communities that we can and must start to embrace diversity and inclusion in real, persistent, and meaningful ways. We wrote this book to educate, inform, inspire, and engage all members of our shared community to embrace diversity and inclusion. The responsibility belongs to us all; it must be shared. We must eagerly take this opportunity to make the necessary changes to ensure innovation in this field for generations to come.

Bibliography

[1] Stasko, J., Choo, J., Han, Y., Hu, M., Pileggi, H., Sadana, R., and Stolper, C. D., Citevis: Exploring conference paper citation data visually, *Posters of IEEE InfoVis*, vol. 2, 2013. xiii

[2] Lorensen, B., On the death of visualization, *Position Papers NIH/NSF Proc. Workshop Visualization Research Challenges*, vol. 1, no. 2, 2004. xiii

[3] Gaither, K., How visualization can foster diversity and inclusion in next-generation science, *IEEE Computer Graphics and Applications*, vol. 37, no. 5, pp. 106–112, 2017. xiv

[4] Bureau of Labor Statistics, National Center for Education Statistics. https://nces.ed.gov 1

[5] Landivar, L. C., Disparities in STEM employment by sex, race, and Hispanic origin, *Education Review*, vol. 29, pp. 911–922, 2013. 1

[6] Dobbs, R., Madgavkar, A., Barton, D., Labaye, E., Manyika, J., Roxburgh, C., et al., The world at work: Jobs, pay, and skills for 3.5 billion people, McKinsey Global Institute Greater, Los Angeles, CA, 2012. 2

[7] Katsomitros, A., *The Global Race for STEM Skills. The Observatory on Borderless Higher Education*, 2013. 2

[8] Grupp, H., *Foundations of the Economics of Innovation*, Edward Elgar Publishing, 1998. 2

[9] Page, S. E., *The Difference: How the Power of Diversity Creates Better Groups, Firms, Schools, and Societies*, Princeton University Press, 2008. 2, 15

[10] Gibbs, K. Jr., Diversity in STEM: What it is and why it matters, *Scientific American*, September 10, 2014. https://blogs.scientificamerican.com/voices/diversity-in-stem-what-it-is-and-why-it-matters/ 2

[11] National Center for Women and Information Technology. https://www.ncwit.org 2

[12] Barnosky, A. D., Ehrlich, P. R., and Hadly, E. A., *Avoiding Collapse: Grand Challenges for Science and Society to Solve by 2050*. Elementa, University of California Press, 2016. http://elementa.ubiquitypress.com/articles/10.12952/journal.elementa.000094/?toggle_hypothesis=on 3

[13] Gregory, R. L., *Eye and Brain: The Psychology of Seeing*, Princeton University Press, 2015. 3

[14] Smith, K., Ed., *Handbook of Visual Communication: Theory, Methods, and Media*, Routledge Press, 2005. 3

[15] Clegg, E. and DeVarco, B., What is the shape of thought? *Shape of Thought*, 2016. shapeofthought.typepad.com 3

[16] Cairo, A., (2016) *A Truthful Art: Why Visualization Will Become an Universal Language*, 2016. https://www.brighttalk.com/webcast/9061/187621/a-truthful-art-why-visualization-will-become-an-universal-language 3

[17] Marrugan, A. E., *Measuring Biological Diversity*, John Wiley & Sons, 2013. 4

[18] Roberson, Q. M., Disentangling the meanings of diversity and inclusion in organizations, *Group and Organization Management*, vol. 31, no. 2, pp. 212–236, 2006. 4, 5

[19] Thomas, D. A. and Ely, R. J., Making differences matter, *Harvard Business Review*, vol. 74, no. 5, pp. 79–90, 1996. 4

[20] U.S. Department of Commerce and Vice President Al Gore's National Partnership for Reinventing Benchmarking Study, Diversity Task Force, 2001. Best practices in achieving workforce diversity. http://govinfo.library.unt.edu/npr/initiati/benchmk/workforce-diversity.pdf 4

[21] https://www.ncwit.org/resources/start-small-start-now-seven-bias-interrupters-male-allies-or-anyone-really-can-start-usi-0 30

[22] Foucault, M., *Discipline and Punish: The Birth of the Prison*, Vintage, 2012. 9

[23] Foucault, M., *The History of Sexuality: The Use of Pleasure*, vol. 2, Vintage, 2012. 10

[24] Frye, M., The politics of reality: Essays in feminist theory, *Crossing Press Feminist Series*, 1983. 11

[25] Young, I. M., In five faces of oppression, *The Community Development Reader*, pp. 346–355, Routledge, 2013. 10

[26] Lloyd, G., The man of reason. Women, knowledge, and reality: Explorations in feminist philosophy, *Psychology Press*, pp. 149–165, 1996. 11

[27] Entman, R. M. and Gross, K. A., Race to judgment: Stereotyping media and criminal defendants, *Law and Contemporary Problems*, 71(4), pp. 93–133, 2008. 12

[28] McIntosh, P., White privilege: Unpacking the invisible knapsack, first appeared in *Peace and Freedom Magazine*, July/August, 1989, pp. 10–12, a publication of the Women's International League for Peace and Freedom, Philadelphia, PA. 12

[29] Crenshaw, K., Demarginalizing the intersection of race and sex: A black feminist critique of antidiscrimination doctrine, feminist theory and antiracist politics. University of Chicago Legal Forum, p. 139, 1989. 12

[30] Leaper, C., Do I belong?: Gender, peer groups, and STEM achievement, *International Journal of Gender, Science and Technology*, 7(2), pp. 166–179, 2015. 13

[31] Rosser, S. V., Breaking into the lab: Engineering progress for women in science, NYU Press, 2012. 13

[32] Johnson, D. R., Women of color in science, technology, engineering, and mathematics (STEM), *New Directions for Institutional Research*, (152), pp. 75–85, 2011. 13

[33] Funk, C. and Parker, K., Women and men in STEM often at odds over workplace equity, *Pew Social Trends*, 2018. 13

[34] Greenwald, A. G. and Banaji, M. R., Implicit social cognition: Attitudes, self-esteem, and stereotypes, *Psychological Review*, 102(1), 4, 1995. 13

[35] Godsil, R. D., Tropp, L. R., Goff, P. A., and Powell, J. A., Addressing implicit bias, racial anxiety, and stereotype threat in education and health care, *The Science of Equality*, 1, 2014. 14

[36] Iliff, R., *How to Champion Inclusion and Diversity within your Business*, Mashable, 2016. `https://mashable.com/2016/05/25/diverse-inclusive-company-cultur e/#Ywcer5LsaZqi` 21

[37] Kerby, S. and Crosby, B., *A Diverse Workforce is Integral to a Strong Economy*, Mashable, 2012. `https://www.americanprogress.org/issues/economy/news/2012/07/12/ 11900/the-top-10-economic-facts-of-diversity-in-the-workplace/` 15

[38] Shin, H. Y. and Park, H. J., What are the key factors in managing diversity and inclusion successfully in large international organizations?, Cornell University, ILR, 2013. `http: //digitalcommons.ilr.cornell.edu/student/45/` 16

[39] Choy, W. K. W., Globalisation and workforce diversity: HRM implications for multinational corporations in Singapore, *Singapore Management Review*, 29(2), 2007.

[40] Sanders, M. G., Overcoming obstacles: Academic achievement as a response to racism and discrimination, *The Journal of Negro Education*, 66(1), pp. 83–93, 1997. 18

[41] Scott, A. and Martin, A., Perceived barriers to higher education in STEM among high-achieving underrepresented high school students of color, *Journal of Women and Minorities in Science and Engineering*, 20(3), pp. 235–256, 2014. 17

[42] The College Board, The 8th annual AP report to the nation, 2012. `http://apreport.c ollegeboard.org/` 18

[43] Darling-Hammond, L., Inequality and the right to learn: Access to qualified teachers in California's public schools, *Teachers College Record*, 106(10), pp. 1936–1966, 2004.

[44] EdTrust-West, The cruel divide: How California's education finance system shortchanges its poorest school districts, 2012. `http://www.edtrust.org/sites/edtrust.org/fi les/ETW%20Cruel%20Dividem%20Report.pdf`

[45] Educational Testing Service (ETS), Access to success: Patterns of advanced placement participation in U.S. high schools, 2008. `http://www.ets.org/Media/Research/pdf /PIC-ACCESS.pdf`

[46] Goode, J., Mind the gap: The digital dimension of college access, *Journal of Higher Education*, 81(5), pp. 583–618, 2010.

[47] Margolis, J., *Stuck in the Shallow End: Education, Race, and Computing*, MIT Press, Cambridge, MA, 2008.

[48] WestEd center for the future of teaching and learning, High hopes-few opportunities: The status of elementary science education in California, 2011. `http://www.scribd.com/doc/70262940/High-Hopes-Few-Opportunities- by-Center-for-the-Future-of-Teaching-and-Learning-at-WestEd#archive` 18

[49] American Association for the Advancement of Science (AAAS), In pursuit of a diverse science, technology, engineering, and mathematics workforce: Recommended research priorities to enhance participation by underrepresented minorities, 2001. `http://ehrw eb.aaas.org/mge/Reports/Report1/AGEP/` 18

[50] Price, J., The effect of instructor race and gender on student persistence in STEM fields, *Economics of Education Review*, 29(6), pp. 901–910, 2010. 18

[51] Chang, M., Eagan, M., Lin, M., and Hurtado, S., Considering the impact of racial stigmas and science identity: Persistence among biomedical and behavioral science aspirants, *Journal of Higher Education*, 82(5), pp. 564–596, 2011. 18

[52] Perna, L., Lundy-Wagner, V., Drezner, N., Gasman, M., Yoon, S., Bose, E., and Gary, S., The contribution of HBCUS to the preparation of African American Women for STEM careers: A case study, *Research in Higher Education*, 50(1), pp. 1–23, 2009.

[53] Thiry, H., Laursen, S., and Hunter, A., What experiences help students become scientists? A comparative study of research and other sources of personal and professional gains for STEM undergraduates, *Journal of Higher Education*, 82(4), pp. 357–388, 2011. 18

[54] Major, B. and O'Brien, L., The social psychology of stigma, *Annual Review of Psychology*, 56, pp. 393–421, 2005. 18

[55] Wong, C., Eccles, J., and Sameroff, A., The influence of ethnic discrimination and ethnic identification on African American Adolescents' School and socioemotional adjustment, *Journal of Personality*, 71, pp. 1197–1232, 2003. 18

[56] Crocker, J., Major, B., and Steele, C., Social stigma, in D. Gilbert, S. Fiske, and G. Lindzey, Eds., *The Handbook of Social Psychology*, 4th ed., 2, pp. 504–553, McGraw-Hill, Boston, MA, 1998. 18

[57] Steele, J., James, J., and Barnett, R., Learning in a man's world: Examining the perceptions of undergraduate women in male-dominated academic areas, *Psychology of Women Quarterly*, 26, pp. 46–50, 2002. 18

[58] Murphy, M., Steele, C., and Gross, J., Signaling threat: How situational cues affect women in math, science, and engineering settings, *Psychological Science*, 18, pp. 879–885, 2007.

[59] Shapiro, J. R. and Williams, A. M., The role of stereotype threats in undermining girls' and women's performance and interest in STEM fields, *Sex Roles: A Journal of Research*, 66, pp. 175–183, 2012. 18

[60] McWhirter, E., Perceived barriers to education and career: Ethnic and gender differences, *Journal of Vocational Behavior*, 50, pp. 124–140, 1997. 18

[61] Harrell, S., A multidimensional conceptualization of racism-related stress: Implications for the well-being of people of color, *American Journal of Orthopsychiatry*, 70(1), pp. 42–56, 2000. 18

[62] Monat, A. and Lazarus, R., *Stress and Coping: An Anthology*, Columbia University Press, New York, 1991. 18

[63] Steele, C. and Aronson, J., Stereotype threat and the intellectual test-performance of African-Americans, *Journal of Personality and Social Psychology*, 69(5), pp. 797–811, 1995. 18

[64] Schmader, T., Major, B., and Gramzow, R., Coping with ethnic stereotypes in the academic domain: Perceived injustice and psychological disengagement, *Journal of Social Issues*, 57(1), pp. 93–111, 2001. 18

[65] Garbee, E., *The Problem with the "Pipeline"*, 2017. `http://www.slate.com/articles/technology/future_tense/2017/10/the_problem_with_the_pipeline_metaphor_in_stem_education.html` 19

[66] Garbee, E., The value of a STEM, Ph.D. thesis, Arizona State University, May 2018. 19

[67] Allen-Ramdial, S. and Campbell, A., Reimagining the pipeline: Advancing STEM diversity, persistence, and success, *Bioscience*, 64(7):612–618, 2014. 19, 20

[68] Cohen, G. L. and Garcia, J., Identity, belonging, and achievement: A model, interventions, implications, *Current Directions in Psychological Science*, 17, pp. 365–369, 2008. 19, 20

[69] Purdie-Vaughns, V., Steele, C. M., Davies, P. G., Ditlmann, R., and Crosby, J. R., Social identity contingencies: How diversity cues signal threat or safety for African Americans in mainstream institutions, *Journal of Personality and Social Psychology*, 94, pp. 615–630, 2008. 19

[70] Cress, C. M. and Sax, L. J., Campus climate issues to consider for the next decade, *New Directions for Institutional Research*, 98, pp. 65–80, 1998. 20

[71] Hirt, J. B. and Muffo, J. A., Graduate students: Institutional climates and disciplinary cultures, *New Directions for Institutional Research*, 98, pp. 17–33, 1998. 20

[72] Hurtado, S., Carter, D. F., and Kardia, D., The climate for diversity: Key issues for institutional self-study, *New Directions for Institutional Research*, 98, pp. 53–63, 1998. 20

[73] Whittaker, J. A. and Montgomery, B. L., Cultivating diversity and competency in STEM: Challenges and remedies for removing virtual barriers to constructing diverse higher education communities of success, *Journal of Undergraduate Neuroscience Education*, 11, pp. A44–A51, 2012. 20

[74] Kendricks, K., Nedunuri, K. V., and Anthony, A. R., Minority student perceptions of the impact of mentoring to enhance academic performance in STEM disciplines, *Journal of STEM Education*, 14(2), pp. 38–46, 2013. 20

[75] Trent, W., Owens-Nicholson, D., Eatman, T. K., Burke, M., and Daugherty, J., Norman K. Justice, equality of educational opportunity, and affirmative action in higher education, in: Chang, M. J., Witt, D., Jones, J., Hakuta, K., Eds., *Compelling Interest: Examining the Evidence on Racial Dynamics in Colleges and Universities*, Stanford University Press, pp. 22–48, 2003. 20

[76] Tanner, K. and Allen, D., Cultural competence in the college biology classroom, *CBE Life Sciences Education*, 6, pp. 251–258, 2007. 20

[77] Campbell, A. G., STEM diversity and critical mass, *IMSD View*, 4(3), 1–2013, April 22, 2014. `http://biomed.brown.edu/imsd/IMSD_View_V4_I3.pdf10.1093/biosci/biu076.html` 20

[78] Steele, C. M., A threat in the air: How stereotypes shape intellectual identity and performance, *American Psychologist*, 52, pp. 613–629, 1997. 20

[79] O'Rourke, S., Diversity and merit: How one university rewards faculty work that promotes equity, *Chronicle of Higher Education*, September 28, 2008. `http://chronicle.com/article/DiversityMerit-How-One/12351` 20

[80] Florentine, S., Diversity and inclusion: 8 best practices for changing your culture, 2018. `https://www.cio.com/article/3262704/hiring-and-staffing/diversity-and-inclusion-8-best-practices-for-changing-your-culture.html` 21

[81] Freiler, C., Building inclusive cities and communities, *Education Canada*, 48(1), pp. 40–42, 2008. 53

[82] Archer, L., Hutchings, M., and Ross, A., *Higher Education and Social Class: Issues of Exclusion and Inclusion*, Routledge, 2005. 53

[83] Chavis, Lee, and Buchanan, *Building Inclusive Communities*, Community Tool Box, 2001. `https://ctb.ku.edu/en/table-of-contents/culture/cultural-competence/inclusive-communities/main` 53

[84] University of Virginia School of Medicine, Establishing a culture of inclusion as a strategy for excellence: A strategic approach, `https://med.virginia.edu/asp/wp-content/uploads/sites/46/2014/04/SOM-Diversity-STRATEGIC-PLAN-6--3-14-2.pdf` 54

[85] Anderson, M. D., *Ten Characteristics of an Inclusive Organization*, 2018. `https://www.mdanderson.org/content/dam/mdanderson/documents/about-md-anderson/careers/diversity-and-inclusion/ten-characteristics-of-an-inclusive-organization.pdf` 54

[86] `https://sc18.supercomputing.org` 56, 58, 59

[87] ECI ethics and compliance initiative, Why have a code of conduct, 2018. `https://www.ethics.org/resources/free-toolkit/code-of-conduct/` 56

[88] Erwin, P. M., Corporate codes of conduct: The effects of code content and quality on ethical performance, *Journal of Business Ethics*, 99, pp. 535–548, 2010. 56

[89] Hardy, M. C., Drafting an effective ethical code of conduct for professional societies: A practical guide, *Administrative Sciences*, 6, 2016. 56

[90] Szczur, K., Building inclusive communities, *Challenging the status quo through ally-ship*, 2017. `https://medium.com/@fox/building-inclusive-communities-232d c01d1aba` 53

[91] Sue, D. W., Capodilupo, C. M., Torino, G. C., Bucceri, J., Holder, A. B., Nadal, K. L., and Esquilin, M., Racial microaggressions in everyday life: Implications for clinical practice, *American Psychologist*, 62(4), pp. 271–286, 2007. 53

[92] `https://cra.org/cra-w/wp-content/uploads/sites/5/2018/05/CRAW-Best-Practices-for-Conferences-v5.pdf` 55, 58, 59

[93] Merton, R. K., *The Sociology of Science: Theoretical and Empirical Investigations*, University of Chicago Press, Chicago, IL, 1973. 57

[94] Le Goues, C., Brun, Y., Apel, S., Berger, E., Khurshid, S., and Smaragdakis, Y., Effectiveness of anonymization in double-blind review, *Communications of the ACM*, 61(6), pp. 30–33, 2018. 58

[95] McKinley, K., More on improving reviewing quality with double-blind reviewing, external review committees, author response, and in person program committee meetings, 2015. `http://www.cs.utexas.edu/users/mckinley/notes/blind.html` 58

[96] Lee, C. J., Sugimoto, C. R., Zhang, G., and Cronin, B., Bias in peer review, *Journal of the American Society for Information Science and Technology*, 64(1), pp. 2–17, 2013. 58

[97] Calisi, R., Opinion: How to tackle the childcare—conference conundrum, *Proc. of the National Academy of Sciences U.S.*, 115(12), pp. 2845–2849, 2018. 59

[98] `http://ieeevis.org/year/2018/info/inclusion-and-diversity/viskids-event` 59

[99] `https://www.kiddiecorp.com` 59

[100] `http://ieeevis.org/year/2018/info/inclusion-and-diversity/diversity-scholarship` 60

[101] `https://www.worldbank.org/en/topic/disability` 60

[102] `http://www.sigaccess.org/welcome-to-sigaccess/resources/accessible-conference-guide/#website` 60

[103] IEEE, IEEE code of ethics, *IEEE Policies, Section 7—Professional Activities (Part A—IEEE Policies)*, 2018. `https://www.ieee.org/about/corporate/governance/p7--8.html` 57

[104] Martin, J. L., Ten simple rules to achieve conference speaker gender balance, *PLoS Computational Biology*, 10(11), 2014. https://journals.plos.org/ploscompbiol/artic le?id=10.1371/journal.pcbi.1003903 60

[105] Byrd, V. and Vieria, C., Visualization: A conduit for collaborative undergraduate research experiences, *Proc. of the ASEE Annual Conference*, 2017.

[106] Barker, T., Byrd emphasizes value of visualization at XSEDE14, *HPCWire*, 2014. https://www.hpcwire.com/2014/07/31/byrd-emphasizes-value-visualization-xsede14/ 31

[107] Byrd, V. and Tanner, L. 1st CRA-W/CDC broadening participation in visualization (BPViz) workshop—a success despite the polar vortex!, *Computing Research News*, 26, 2014. 31

[108] Gregerman, S., Lerner, J. S., von Hippel, W., Jonides, J., and Nagda, B. A., Undergraduate student-faculty research partnerships affect student retention, *The Review of Higher Education*, 22(1), pp. 55–72, 1998. 33

[109] Byrd, V. L. and Vieira, V. C., Visualization: A conduit for collaborative undergraduate research experiences, *American Society for Engineering Education*, 2017. 33

[110] White House Initiative on Historically Black Colleges and Universities School Directory, 2014. https://sites.ed.gov/whhbcu/files/2014/09/HBCU-Directory.pdf 34

[111] Hispanic Association of Colleges and Universities, 2018. https://www.hacu.net/ass nfe/CompanyDirectory.asp?STYLE=2&COMPANY_TYPE=1 34

[112] https://www.whitehouse.gov/sites/whitehouse.gov/files/images/NSCI%20S trategic%20Plan.pdf 35

[113] Raz, A. and Buhle, J., Typologies of attentional networks, *Nature Reviews Neuroscience*, 7(5), pp. 367–379, 2006. 35

[114] Hartley, J. and Davies, I. K., Note-taking: A critical review, *Innovations in Education and Training International*, 15(3), pp. 207–224, 1978.

[115] Shirey, L., Importance, interest, and selective attention, *The Role of Interest in Learning and Development*, pp. 281–296, 1992. 35

[116] Shantz, C., 12 brain/mind learning principles in action: The fieldbook for making connections, teaching, and the human brain, *Teaching Theology and Religion*, 9(3), pp. 189–190, 2006. 35

[117] Cangelosi, P. R. and Whitt, K. J., Teaching through storytelling: An exemplar, *International Journal of Nursing Education Scholarship*, 3(2), 2006. http://dx.doi.org/10.2202/1548--923X.1175 35

[118] McBride, D. M. and Dosher, B. A., A comparison of conscious and automatic memory processes for picture and word stimuli: A process dissociation analysis, *Consciousness and Cognition*, 11(3), pp. 423–460, 2002. 36

[119] Endestad, T., Magnussen, S., and Helstrup, T., Memory for pictures and words following literal and metaphorical decisions, *Imagination, Cognition and Personality*, 23(2), pp. 209–216, 2003.

[120] Stenberg, G., Conceptual and perceptual factors in the picture superiority effect, *European Journal of Cognitive Psychology*, 18(6), pp. 813–847, 2006. 36

[121] Post, C., Lia, E., Ditomaso, N., Tirpak, T., and Borwankar, R., Capitalizing on thought diversity for innovation, *IEEE Engineering Management Review*, 1(39), pp. 100–114, 2011. 36

[122] Networking and information technology research and development program (U.S.), *The Federal Big Data Research and Development Strategic Plan*, 2016. http://books.google.com/books/about/The_Federal_Big_Data_Research_and_Develo.html?hl=&id=5bBDvgAACAAJ 36

[123] Dunbar, R. I. M., The social brain hypothesis, *Evolutionary Anthropology: Issues, News, and Reviews*, 6(5), pp. 178–190, 1998. 36

[124] Flora, K., Bromley, J., and Bracken, A., Using the familiar to teach the unfamiliar: Active learning strategies in research methods. 36

[125] Silberman, M., *Active Learning: 101 Strategies to Teach Any Subject*, Prentice Hall, 1996.

[126] Meyers, C. and Jones, T. B., (Vella, F., 1994), *Promoting Active Learning: Strategies for the College Classroom*, p. 192, Jossey-Bass, San Francisco, 1993. *Biochemical Education*, 22(1), p. 61. 36

[127] https://www.sighpc.org/for-our-community/computing4change/2018-competition-participants 38

[128] Committee on underrepresented groups and the expansion of the science and engineering workforce (U.S.) and committee on science, engineering, and public policy (U.S.) and national research council and others, *Expanding Underrepresented Minority Participation: America's Science and Technology Talent at the Crossroads*, National Academies Press, 2010. 39

[129] Dasgupta, N. and Stout, J. G., Girls and women in science, technology, engineering, and mathematics: STEMing the tide and broadening participation in STEM careers, *Policy Insights from the Behavioral and Brain Sciences*, 1(1), pp. 21–29, 2014. 39

[130] Sorcinelli, M. D. and Yun, J., From mentor to mentoring networks: Mentoring in the new academy, *Change: The Magazine of Higher Learning*, vol. 39, no. 6, pp. 58–61, 2007. 39

[131] DFG, Cooperation with developing countries, *Deutsche Forschungsgemeinschaft*, September 20, 2018. http://www.dfg.de/en/research_funding/programmes/internatio nal_cooperation/developing_countries/index.html 81

[132] Science and engineering indicators, (NSB-2018-1), digest (NSB-2018-2), January 2018. https://nsf.gov/statistics/2018/nsb20181/ 79, 81, 82

[133] Bohannon, J. and Doran, K., Vast set of public CVs reveals the world's most migratory scientists, *Science*, 356(6339), pp. 691–692, May 18, 2017. https://www.sciencemag.org/news/2017/05/vast-set-public-cvs-reveals-world-s-most-migratory-scientists

[134] NIH, Cooperation with developing countries, *National Institutes of Health*, October 9, 2018. https://www.nih.gov/

[135] USAID, U.S. agency for international development, October 4, 2018. https://www.us aid.gov/

[136] Dunbar, B., National Aeronautics and Space Administration, 2018. https://www.nasa .gov/

[137] Smithsonian, Smithsonian Institute, 2018. https://www.si.edu/

[138] USDA, U.S. Department of Agriculture, 2018. https://www.usda.gov/

[139] USGS, U.S. Geological Services, 2018. https://www.usgs.gov/

[140] The National Academies of Sciences, Engineering, and Medicine, T. 202.334.2000, 2018. http://sites.nationalacademies.org/pga/peer/index.htm 80

[141] Borg, A., 2018. https://anitab.org/ 80

[142] WomenTechmakers, Scholars program, 2018. https://www.womentechmakers.com/ scholars 80

[143] ACM-W, Supporting, celebrating and advocating for women in computing, 2018. http s://women.acm.org/ 80

[144] Holly Else, EU to world: Join our €100-billion research programme European Commission, *Nature*, 2018. https://www.nature.com/articles/d41586-018-05392-7 81

[145] ACM-W Europe, ACM Europe women program, 2018. http://europe.acm.org/acm-w-europe.html

[146] EPWS, European Platform of Women Scientists, 2018. https://epws.org/

[147] DFG, The research-oriented standards on gender equality, October 2, 2018. http://www.dfg.de/en/research_funding/principles_dfg_funding/equal_opportunities/research_oriented/ 81

[148] EPSRC, EPSRC funding for international collaboration, 2018. https://epsrc.ukri.org/funding/applicationprocess/routes/international/ 81

[149] UNESCO, World day for cultural diversity for dialogue and development, 2018. https://en.unesco.org/commemorations/culturaldiversityday/2018 81, 82

[150] UNESCO, Supporting women scientists: Mentoring, networks and role models, *Gender and Science*, 2017. http://www.unesco.org/new/en/natural-sciences/priority-areas/gender-and-science/supporting-women-scientists/

[151] Schmidt, S., The national interest, *Saudi Money Shaping U.S. Research*, February 11, 2013. https://nationalinterest.org/commentary/saudi-money-shaping-us-research-8083 81

[152] Mertl, M., Sciencemag, *Grants for Women in Science*, May 31, 2000. http://www.sciencemag.org/careers/2000/05/grants-women-science 83

[153] fulbrightonline, Fulbright U.S. student program, 2018. https://us.fulbrightonline.org/#&panel1--1 83

[154] IASTE United States, 2018. http://www.iasteunitedstates.org/ 83

[155] Isenberg, P., Isenberg, T., Sedlmair, M., Chen, J., and Möller, T., Visualization as seen through its research paper keywords, *IEEE Transactions on Visualization and Computer Graphics*, 23(1):771–780, January 2017. 65, 66, 67, 70

[156] Isenberg, P., Heimerl, F., Koch, S., Isenberg, T., Xu, P., Stolper, C., Sedlmair, M., Chen, J., Möller, T., and Stasko, J., Visualization publication dataset, *Dataset and Web Page*, 2015. 65

[157] Isenberg, P., Isenberg, T., Sedlmair, M., Chen, J., and Möller, T., KeyVis, *Online Database*, 2014. http://keyvis.org/

[158] Isenberg, P., Isenberg, T., Sedlmair, M., Chen, J., and Möller, T., Toward a deeper understanding of visualization through keyword analysis, *Research Report RR-8580*, INRIA, August 2014. Also published on arXiv.org (# 1408.3297). 67

[159] Frida, Regional fund for digital innovation in Latin America and the Caribbean, 2018. `https://programafrida.net/en/` 82

[160] Abel, G. J. and Sander, N., science.sciencemag, *Quantifying Global International Migration Flows*, vol. 343, issue 6178, pp. 1520–1522, March 28, 2014. `http://science.sciencemag.org/content/343/6178/1520` 82

[161] The World Bank, 2018. `https://datacatalog.worldbank.org/dataset/world-development-indicators`

[162] *Proc. of the 1st IEEE Conference on Visualization*, (Cat. No.90CH2914–0), p. 1, San Francisco, CA, 1990. `http://ieeexplore.ieee.org/stamp/stamp.jsp?tp=&arnumber=146352&isnumber=3914` 65

[163] Horizon 2020, November, 2018. `https://ec.europa.eu/programmes/horizon2020/en/` 81

[164] Bohannon, J. and Doran, K., Introducing ORCID, *Science*, 356(6339), pp. 691–692, 2017. `https://doi.org/10.1126/science.356.6339.691` 83

[165] Vast set of public CVs reveals the world's most migratory scientists, *People and Event, Scientific Community, Social Sciences. Science Magazine.* `http://www.sciencemag.org/news/2017/05/vast-set-public-cvs-reveals-worlds-most-migratory-scientists` 83

[166] €100-billion budget proposed for Europe's next big research programme, *Nature*, 557(150), 2018. `https://www.nature.com/articles/d41586--018-05105-0` 81

[167] The Africa-EU partnership, November 2018. `https://www.africa-eu-partnership.org/en` 81

[168] Grupp, H., *Foundations of the Economics of Innovation*, Books, Edward Elgar Publishing, no. 1390, 1998. `https://ideas.repec.org/b/elg/eebook/1390.html` 81

[169] Mahroum, S., Highly skilled globetrotters: Mapping the international migration of human capital, *R&D Management*, 30(1), pp.23–32, 2000. `https://doi.org/10.1111/1467--9310.00154` 83

[170] Yau, N., Most female and male occupations since 1950, 2018. `https://flowingdata.com/2017/09/11/most-female-and-male-occupations-since-1950/` 70

[171] Holland, J. H., Genetic algorithms, *Scientific American*, 267(1), pp. 66–73, 1992. 63

[172] Haslett, B. B. and Lipman, S., *Micro Inequities: Up Close and Personal*, Sage Publications, Inc., 1997. 25

Authors' Biographies

EDITORS

RON METOYER

Ron Metoyer is an Associate Professor of Computer Science and Engineering at the University of Notre Dame. He earned his B.S. in Computer Science and Engineering at the University of California, Los Angeles (1994) and his Ph.D. in Computer Science from the Georgia Institute of Technology (2002) where he worked in the Graphics, Visualization and Usability Center. He previously served on the faculty at Oregon State University in the School of Electrical Engineering and Computer Science (2001–2014). His primary research interest is in human-computer interaction and information visualization, with a focus on multivariate data visualization, decision-making, and narrative. He has published over 60 papers in top conferences and journals in human-computer interaction and computer graphics, and he is the recipient of a 2002 NSF CAREER Award for his work in exploring usability issues around the generation of animated character content for training scenarios. As an advocate for broadening participation in computing, he has served in many roles including several years on the program committee of the Richard Tapia Celebration of Diversity in Computing. He also serves as Assistant Dean in the College of Engineering at the University of Notre Dame.

KELLY GAITHER

Kelly Gaither is the Director of Visualization at the Texas Advanced Computing Center (TACC) at the University of Texas, having joined TACC as Associate Director in September 2001. She received her doctoral degree in Computational Engineering from Mississippi State University in May 2000 and her masters and bachelor's degrees in Computer Science from Texas A&M University in 1992 and 1988, respectively. She has given a number of invited talks and published over thirty refereed papers in fields ranging from computational mechanics to supercomputing applications to scientific visualization. Over the past ten years, she has actively participated in conferences related to her field, specifically acting as general chair of IEEE Visualization in 2004.

CONTRIBUTORS

ALFIE ABDUL-RAHMAN

Alfie Abdul-Rahman is a Lecturer in Computer Science at King's College London. She received her Ph.D. in computer science from Swansea University. Before joining King's College London, she was a Research Associate at the University of Oxford e-Research Center. She worked on document engineering as a Research Engineer in HP Labs Bristol and then on multi-format publishing as a software developer in London. Her research interests include visualization, computer graphics, and human-computer interaction.

ANASTASIA BEZERIANOS

Anastasia Bezerianos is an assistant professor at the University of Paris-Sud and a member of the ILDA Inria team, in France. She received a M.Sc. and Ph.D. from the University of Toronto, studying interaction designs for large displays. Her work is at the intersection of human-computer interaction and information visualization, and she serves regularly on conferences program committees for both HCI (CHI, UIST) and visualization (VIS). She is also interested in interaction and visualization designs for very large and shared displays, particularly in collaborative settings, as well as aspects related to the evaluation of interactive visualizations and their use to support data analysis and decision making. She served as Community co-chair in IEEE VIS for two years (2017–2018) and will be one of its Ombuds chairs in 2019.

RITA BORGO

Rita Borgo is a Senior Lecturer in the Informatics Department at King's College London (KCL) and current Head of the Human Centered Computing research group. Her main research interests lie in the areas of information visualization and visual analytics, with particular focus on the role of human factors in visualization. Her research has followed an ambitious program of developing new data visualization techniques for interactive rendering and manipulation of large multi-dimensional and multivariate datasets. Novel in all aspects of her research is the aim to provide solutions that involve humans in the loop of intelligent reasoning while reducing the burden of inspection of large complex data. Her research has been awarded support from the Royal Society, EPSRC, and EU. She is currently championing the newly created Urban Living hub at KCL and works in close collaboration with the Center for Urban Science and Progress (CUSP)—London to the increase impact of visualization within urban-related challenges.

MICHELLE BORKIN

Michelle Borkin is an Assistant Professor in the Khoury College of Computer Sciences at Northeastern University and works on the development of novel visualization techniques and tools to enable new insights and discoveries in data. She works across disciplines to bring together computer scientists, doctors, and astronomers to collaborate on new analysis and visualization techniques and cross-fertilize techniques across disciplines. Her research resulted in the development of novel computer-assisted diagnostics in cardiology and radiology, scalable visualization solutions for large network data sets, and novel astrophysical visualization tools and discoveries. Prior to joining Northeastern, Professor Borkin was a postdoctoral research fellow in computer science at the University of British Columbia as well as an associate in computer science at Harvard and a research fellow at Brigham & Women's Hospital. She received her Ph.D. in Applied Physics at Harvard's School of Engineering and Applied Sciences (SEAS) in 2014. She also earned an M.S. in Applied Physics and a B.A. in Astronomy and Astrophysics & Physics from Harvard University. She was previously a National Science Foundation (NSF) graduate research fellow, a National Defense Science and Engineering graduate (NDSEG) fellow, and a TED fellow.

VETRIA BYRD

Vetria Byrd is an Assistant Professor of Computer Graphics Technology and Director of the Byrd Data Visualization Lab in the Polytechnic Institute at Purdue University's main campus in West Lafayette, Indiana. Dr. Byrd is introducing and integrating visualization capacity building into the undergraduate data visualization curriculum. She is the founder of the Broadening Participation in Visualization (BPViz) Workshop and served as a steering committee member on the Midwest Big Data Hub (2016–2018). She has taught data visualization courses on national and international platforms as an invited lecturer of the International High-Performance Computing Summer School (IHPCSS). Her visualization webinars on Blue Waters, a petascale supercomputer at the National Center for Supercomputing Applications at the University of Illinois at Urbana-Champaign, introduce data visualization to audiences around the world. As described in her invited plenary talk featured on *HPC Wire*, Dr. Byrd utilizes data visualization as a catalyst for communication, a conduit for collaboration, and a platform to broaden participation of underrepresented groups in data visualization. Dr. Byrd's research interests include data visualization, data analytics, data integration, visualizing heterogeneous data, and the science of learning and incorporating data visualization at the curriculum level and everyday practice. Dr. Byrd is currently serving as co-Vice Chair of the 2019 Gordon Research Conference on Visualization in Science and Education: Educating Skillful Visualizers.

ALEXANDRA DIEHL

Alexandra Diehl is a postdoctoral researcher at the Data Analysis and Visualization group in Konstanz. She is a specialist in the discipline of visualization, particularly visual analytics of geospatial data. Alexandra holds a Ph.D. in Computer Science from the University of Buenos Aires, Argentina. She has been working in the area of scientific visualization since 2006, especially oriented toward geospatial data visualization since 2008. Her main interests center on visualization of big data, with a focus on geospatial data, environmental phenomena, and human behavior.

AVIVA FRANK

Aviva Frank will receive her B.A. in Sociology and Gender Studies with a minor in Political Science from SUNY Purchase in the spring of 2019 and will enter a sociological research graduate program at Columbia University in the fall. She is currently conducting research at the intersection of sociology and gender studies, focusing on LGBTQ sexuality and identity formation. She is passionate about social justice and inequality studies, and she actively works to enrich her activism through her studies, and her studies through her activism. In her free time, Aviva enjoys going to museums and botanical gardens, petting animals, and discussing critical theory.

MASHHUDA GLENCROSS

Dr. Mashhuda Glencross is director of research at Pismo Software. She has lectured at Leeds and Loughborough Universities, and worked as a product manager at ARM Ltd. and as a Research Fellow at the University of Manchester. Her work has spanned a diverse range of areas within graphics and visualization including 3D reconstruction/imaging, material acquisition/modeling, visual perception, real-time/massive-model rendering, virtual reality, and haptics. She is currently an elected Director-at-Large for ACM SIGGRAPH.

PETRA ISENBERG

Petra Isenberg is a research scientist (CR) at Inria, Saclay, France, in the Aviz research group. Prior to joining Inria, she received her Ph.D. from the University of Calgary in 2010, working with Sheelagh Carpendale on collaborative information visualization. Petra also holds a Diplom-Ingenieur degree in Computational Visualiztics from the University of Magdeburg. Her main research areas are information visualization and visual analytics, with a focus on off-desktop data analysis, interaction, and evaluation. She is particularly interested in exploring how people can most effectively work together when analyzing data sets on novel display technology such as smart watches, wall displays, or tabletops. Petra has contributed over 70 publications in the top venues in the fields of visualization and HCI, including one best paper and 6 best paper honorable mention awards. She is associate editor-in-chief at *IEEE CG&A*, associate editor of the *IEEE Transactions on Visualization and Computer Graphics*; has served on many organizing committee roles at IEEE VIS, including as papers co-chair for Information Visualization (InfoVis); and has been the co-chair of the biennial Beliv workshop since 2012.

MANUEL A. PÉREZ QUIÑONES

Manuel A. Pérez Quiñones is Associate Dean of the College of Computing and Informatics and Professor of Software and Information Systems at the University of North Carolina at Charlotte (UNCC). His research interests include personal information management, human-computer interaction, computer science education, and diversity issues in computing. He holds a D.Sc. from The George Washington University and a B.A. & M.S. from Ball State University. He has published over 100 refereed articles. Before joining UNCC, he worked at Virginia Tech and the University of Puerto Rico-Mayaguez, as a Visiting Professor at the US Naval Academy, and as a Computer Scientist at the Naval Research Lab. In 2011, he was recognized with a College of Engineering Dean's Award for Excellence in Service (Virginia Tech). In 2017, he received the Richard A. Tapia Achievement Award for Scientific Scholarship, Civic Science and Diversifying Computing. In 2018, he was awarded the CRA Nico A. Haberman award for his contributions aimed at increasing the numbers and/or successes of underrepresented members in the computing research community. He is originally from San Juan, Puerto Rico.

MEG PIRRUNG

Meg Pirrung is a Senior User Experience Designer at the Pacific Northwest National Lab. She earned a Ph.D. in Computational Bioscience from the University of Colorado Anschutz in 2015, and a B.S. in Computer Science from Bowling Green State University in 2008. Her project domains include Machine Learning, Deep Learning, explainable AI, big data visualization, interaction research, streaming data interaction, social media visual analytics, and bioinformatics.

ANNIE PRESTON

Annie Preston is a fifth-year Ph.D. candidate in the Visualization and Interface Design Innovation (VIDI) group at the University of California, Davis. Previously, she received a B.S. in Physics and Astronomy from Haverford College in Pennsylvania. Her research interests include creating visualization tools for scientists and others to better understand uncertainty in data and to reveal the ambiguities inherent to data analysis and modeling.

BERNICE ROGOWITZ

Bernice Rogowitz earned her Ph.D. in Experimental Psychology at Columbia University and was a post-doctoral fellow in Sensory Psychophysics at Harvard University. Her research explores perceptual and cognitive topics related to the representation and exploration of data, including color and color scales, image semantics, shape perception and haptic interfaces. Her main experience in diversity comes through her collaborations with analysts in finance, medicine, business, and science to build complex visualization applications. She inaugurated the IS&T Women in Electronic Imaging group and regularly mentors young professional women and students from non-American cultures. Dr. Rogowitz is the Chief Scientist at Visual Perspectives, a research and consulting practice; teaches at Columbia University, where she designed the Data Visualization course for the SPS Masters' Program in Applied Analytics; and is the founding co-Editor-in-Chief of the new multidisciplinary *Journal of Perceptual Imaging*.

JOHANNA SCHMIDT

Johanna Schmidt is a researcher in visualization and visual analytics. In 2016, she received her Ph.D. from TU Wien, where she especially concentrated on the comparative visualization of large and complex datasets. Afterward, she worked as a scientist at the AIT Austrian Institute of Technology GmbH, Center for Mobility Systems in Vienna, Austria, where she was involved in applied research projects with industry partners. Her current research focuses on the visual analysis of mobility and movement data as well as on the comparative visualization of multiple heterogeneous movement and trajectory data sets. Further research interests include visualization for data science and new perceptual measures for the evaluation of visualizations. She regularly publishes papers in conference proceedings and journals related to visualization and organized a workshop on summarization and report generation at the VIS 2018 conference. She currently serves as a co-chair of IEEE Women in Engineering in Austria.

JONATHAN WOODRING

Jonathan Woodring is a research scientist at the Los Alamos National Laboratory (LANL). He specializes in visualization, scientific data analysis, and high-performance computing. Jon has mentored many graduate students over the years at LANL in collaboration with universities, and he actively participates in outreach and mentoring programs.